Agri Startups for Smart Farming

Agri Startups for Smart Farming

S. Ayyappan

Mukund A. Kataktalware

Letha Devi G.

International Books & Periodical Supply Service

(Publisher of Scientific Books)

38, NISHANT KUNJ, PITAM PURA MAIN ROAD

DELHI-110034 (India)

Phone : 011-27352078, Mobile : 09810146811, E-mail : hkjain1975@yahoo.com

Published by :

International Books & Periodical Supply Service

(Publisher of Scientific Books)
38, NISHANT KUNJ, PITAM PURA MAIN ROAD
DELHI-110034 (India)
Phone : 011-27352078, Mobile : 09810146811, E-mail : hkjain1975@yahoo.com

ISBN 978-93-88892-01-8

Composed, Designed & Printed in India

Preface

Agriculture in India is a way of life. Along with multiple dimensions of food production systems, it has impacted livelihoods as well as other segments of economy. In fact, the reference points for the Indian financial management are the arrival of the annual monsoon as well as the market arrivals. Technologies and practices are transforming every sphere of economy like never before. Similarly, farming is getting impacted by the changing business approaches. Efficient agriculture and speciality agriculture with market connectivity are the new dimensions being discussed and incorporated. The paradigm shift required from production agriculture to secondary agriculture for remunerative agriculture has by now been fully appreciated.

In the 'seed-to-market' approach for profitable agriculture, innovations are the key to reduce efficiency losses at various levels. Shift from subsistence farming to enterprise, agriculture is the future of food production systems. It involves both technical as well as entrepreneurial inputs to make farming 'smart'. The concepts of agri-incubators and startups are finding their roots in the recent past, with partners from all over, even outside agriculture. These have brought in new ways of thinking and conducting farm operations, both on farm and off farm. These also require resource assessment, strategic planning, skill development, risk assessment, communication and management as well as financial and policy environment.

It is in this context a National Consultation on 'Agri Startups for Smart Farming' was organized to analyse the functioning of the incubators and start ups, discuss the high points and constraints, and plan a way forward in terms of finance and policy imperatives in the context of emerging secondary and corporate agriculture. We are grateful to Dr H K Bhanwala, Chairman, NABARD, Mumbai; Dr H R Dave, Deputy Managing Director, NABARD, Mumbai; Dr U S Saha, MD, NABFINS; Dr P V S Surya Kumar, Chief General Manager, NABARD, Bengaluru; Dr Raghavendra Bhatta, Director, ICAR-NIANP, Bengaluru; Dr M J Chandre Gowda, Director, ICAR-ATARI, Bengaluru and Dr K P Ramesha, Head, ICAR- NDRI SRS, Bengaluru; for the guidance and facilitation for the organization of National Consultation. Special gratitude is placed on record to TATA TRUSTS and Social Alpha, for their kind support and co-operation. Contributions of team of convenors; Dr M C A Devi, Dr S Subhash, Dr P K Dixit, Dr M H Sathish Kumar, Dr B Srinivas, Dr S Anandan,

Dr N K S Gowda, Dr S Senani, Dr Menon Rekha Ravindra, Dr Atul Kolte, Dr P K Malik, Dr K Giridhar, Dr P Heartwin Amaladhas, Dr F Magdaline Eljeeva Emerald, Dr A Arangasamy, Dr M Sivaram, Dr K Binsila and Dr K Jayaraja Rao are gratefully acknowledged. Assistance rendered by Ms Kavyashree, M C, Ms K P Anjana, Dr Niketha L and Mr Manoj M is deeply appreciated. Thanks are due to all the authors for enriching this publication with their intellectual inputs. We are sure, this book will bring greater clarity for making 'beyond farm' approach an integrated component of successful agriculture in India and startups make farming a 'win-win-win' for producers, consumers as well as themselves.

16 February, 2019 **Editors**

About the Editors

S. Ayyappan has significantly contributed to the R&D of fisheries sector in particular and agriculture in general. He has served the country in different roles of Director, ICAR-CIFE, Mumbai; DDG (Fisheries), ICAR, New Delhi and Secretary, DARE and Director General, ICAR before joining as the NABARD Chair Professor at ICAR-NDRI, SRS, Bengaluru. His scientific contributions towards the development of agriculture are well appreciated and he has been bestowed with several recognitions and awards. Dr. Ayyappan is presently serving as Chancellor to Central Agricultural University, Imphal.

Mukund A. Kataktalware did his graduation in Veterinary Sciences and Animal Husbandry from Veterinary College, Nagpur in 1997. He completed her Master's and Doctoral degree in Livestock Production and Management from National Dairy Research Institute in 2001 and 2004, respectively. Dr. Kataktalware joined Indian Council of Agricultural Research as a scientist in 2004 at National Research Centre on Yaks, Dirang, Arunachal Pradesh, where he served for about five years. Presently Dr. Kataktalware is working as a Senior Scientist under ICAR-National Dairy Research Institute, Southern Regional Station, Bangaluru. He later did his PG Diploma in Intellectual Property Rights from IGNOU, New Delhi in 2013. He has published a number of research papers and popular articles in various national and international journals of repute. He has published a book on Scientific Dairy Farming and successfully guided several PG students for their research work.

Letha Devi G. did her graduation in Agriculture from Kerala Agricultural University in 2001 and topped the university. She completed her Master's in Dairy Extension from National Dairy Research Institute in 2004 with Director's Gold Medal. She completed her doctoral degree from Indian Agricultural Research Institute, New Delhi in the discipline of Agricultural Extension in the year 2007. She joined Indian Council of Agricultural Research as a Scientist in 2007. She is working as a Scientist under ICAR-National Institute of Animal Nutrition and Physiology, Bangaluru. She did PG Diploma in Intellectual Property Rights from IGNOU, New Delhi in 2013. She has published a number of research papers and popular articles in various national and international journals of repute.

Contents

Chapter **1**

Technology Backstopping to Promote Startups Ecosystem in India - A Case Study

Razia Mohammed A., George Ninan, Ravishankar C.N.
ICAR - Central Institute of Fisheries Technology, Cochin, Kerala
E-mail: cnrs2000@gmail.com

India is fast emerging as a startup nation and technology landscape has witnessed a tremendous growth towards creation of innovative startups and has emerged as the 3rd fastest growing hub for technology startups in the country. As per the NASSCOM start-up report, 2018, India has gained the position as a global startup hub that is becoming attractive for investors, startups and corporates. Innovative and novel technologies are increasingly recognized as one of the crucial factors to enhance the business' competitiveness and gain success in the global market-driven economy.

With the aim of enhancing the foundation for new technology based industries and developing a knowledge-based economy, particularly in the fisheries sector, ICAR-Central Institute of Fisheries Technology (ICAR-CIFT) started an Agri-Business Incubation (ABI) Centre. This entrepreneurial support system handholds entrepreneurs to shift their economic resources through strong technical and advisory support, into areas that yield higher productivity and returns. It focuses on finding new ways of doing business in fisheries and allied agricultural fields, through wide spectrum of activities. The Centre helps prospective entrepreneurs, by providing pro-active and value-added business support in terms of technology transfer, consultancy, infrastructure facility, experts' guidance and specialized training programmes to develop technology-based business ideas and establish sustainable enterprises. It acts as a platform for the speedy commercialization of the technologies, through an interfacing and networking mechanism between research institutions, industries and financial institutions. The incubator at ICAR-CIFT differs from traditional Business Incubators as it is tailored specifically for technology-based industries and is operational at an area with a high concentration of fish production. This industry-specific incubator also allows new firms to tap into scientific knowledge and business networks. ABI offers their services to industries not only in Cochin, but also all over the world through virtual incubation.

Observations/ Remarks

Indian Agritech startups are innovating across the agricultural value chain from weather prediction and sowing advisories to farm output through web portals. As per the NASSCOM reports, 225+ new Agritech startups incorporated between 2015 and 2016. While a number of new startups have been moderated since then, but annual investments have gone up 25 times demonstrating investor focus on mature Agritech startups. Agritech startups raised $313mn in 2016 funding deals seen in 2017/2018 primarily focused on IoT and market linkage platforms. Government and academia continue to provide nurturing environments in the form of incubators by offering mentoring, incentives, event participation opportunities, etc.

With more than 350 startups and over $300mn funding in 2016, Indian Agritech sector has witnessed a sustained growth during the past two years. Digital technologies and startups in the field of supply chain, farm data analytics, infrastructure and information platforms are favourite among investors. AI based smart solutions, advanced farm analytics, and increased usage of new technologies like drones would define the next growth phase of the industry.

The Centre launched the Startup India Policy in January 2016 with the aim of building a strong ecosystem to promote startups and entrepreneurs across the country. While the startup narrative was largely dictated by cities, such as Bengaluru, Delhi-NCR, Mumbai, Chennai, Hyderabad and Pune, startups are now emerging in smaller cities too, including unlikely, such as Shillong, Imphal, Srinagar and Panjim. The Department of Industrial Policy and Promotion (DIPP) reveals the total number of startups registered with the Startup India Hub stands at 2,46,665 across India. While 12,867 have been recognized by the government, 129 have received funding from the government's fund. During the year 2018, a few more States have come out with policies specifically designed to harness the potential of startups. These states have laid out the groundwork in terms of policy framework and the incentives being offered.

Discussions

Globally Agritech has come a long way both, in terms of investment and technology. Increased and timely support to early stage start-ups will boost the sector further in India. As per the NASSCOM reports (2018), funding in the Indian Agritech sector is 10% of Global funding but startups struggle to scale up. There is a need for large companies to effectively collaborate with startups. More than 25% of farmers in India today have access to smart phones. There is a need to create mobile trainings to educate the farmers and help them adapt and adopt to new technological advancements. Only 9% of all funding in last 5 years was focused on growth stage startups. This emphasizes the need for Corporate and Government accelerators to help Agritech startups grow to the next level. Now-a-days, innovation is increasingly recognized as one of the crucial factors to enhance the business's competitiveness and gain success in the global market-driven economy.

According to some credible surveys, it is indicated that firms that know how to manage innovation, can seize new business opportunities, become key players in growth industries and turn themselves into highly profitable companies. It is true not only for big enterprises but also remains the case for small ones as it might serve as a key to differentiate themselves from the rivals. Most of SMEs, especially start-ups are increasingly recognizing the importance and opportunities of innovation in enhancing their competitiveness and accessing to regional and global markets in the context that they enter the markets at later stage and with less advantages and resources.

Agri-Business Incubation Centre at ICAR-CIFT

Fisheries sector with its important role played in the socio-economic development of the country has become a powerful income and employment generator, and stimulates the growth of a number of subsidiary small, medium and large scale industries. In order to translate the research results arising from the field of fisheries and other agricultural sectors, ICAR has set up an innovation based Business Incubation Centre (BIC) at CIFT, Cochin. BIC is managed by ZTM-BPD Unit and aims at establishment of fisheries enterprises through IPR enabled ICAR technologies. This industry-specific incubator also allows new firms to tap into local knowledge and business networks that are already in place. ABI offers their services to industries not only in Cochin, but also all over India through virtual incubation. Beyond promoting business growth, the Centre is also trying to bring its benefits to all the fisheries communities in India.

Through this Centre, the Institute has reached out to more than 150 small and medium-sized technology and technology-enabled enterprises and has helped to create good number of jobs. ICAR-CIFT was able to successfully commercialize many technologies through the ABI Centre, and were the first among all ICAR Business Incubators to accomplish an international technical consultancy. Every month, more than 20 people approach the Centre for seeking business assistance, linkages, advice and tool. There are around 146 registered incubatees at ABI, from all over India and it possess good infrastructure facilities suitable for providing direct incubation to nine entrepreneurs in a corporate environment within the premises of Institute at a time. About thirty firms catering to diverse fields in fisheries sector have already graduated from the Centre.

ABI Centre supports operations on business projects as a measure of enhancing the foundation for new technology-based industries and establishing a knowledge-based economy. It focuses on finding new ways of doing business in fisheries and allied agricultural fields by finding doors to unexplored markets. The Centre helps prospective entrepreneurs, by providing pro-active and value-added business support in terms of technical consultancy, infrastructure facility, experts' guidance and training to develop technology-based business ideas and establish sustainable enterprises. It acts as a platform for the speedy commercialization of the ICAR technologies, through an interfacing and networking mechanism between research institutions, industries and financial institutions. This unique Fisheries Incubator is now known as a "One Stop Shop", where entrepreneurs can receive pro-active, value-added support in terms of technical consultancy, and access to critical tools, such as entrepreneur ready technologies, vast infrastructure and other resources that may otherwise be unaffordable, inaccessible or unknown. With the aim of transforming the incubator into a symbol of entrepreneurship and innovation, the Centre has created an environment for accessing timely scientific and technical assistance and support, required for establishment of technology, based business ventures. The activities of the ABI Centre focuses on finding creative and innovative ways for linking public sector resources and private sector initiatives within and across regional and national boundaries for promoting economic growth. The Centre uses the right expertise in relevant fields to identify and analyze the constraints and barriers hindering the growth of a business, and devise appropriate strategies. It explores the various structures and strategies to help small enterprises to grow and ensure a promising future in the global market. It fosters corporate and community collaborative efforts, while nurturing positive government-research-business relationships.

Process of Business Incubation

ABI Centre targets entrepreneurs, from fledgling startups in need of basic small scale processing capacity to sophisticated businesses in need of R&D back up, office infrastructure and pilot/test market processing facility for the development of new products. It possesses good infrastructure facilities suitable for providing direct incubation of about nine entrepreneurs in a corporate environment within the premises of CIFT, at a time. The purpose of direct incubation is to support emerging companies through their infancy. BIC apart from being a multi-tenant facility with on-site management that delivers an array of entrepreneurial services to clients operating with the facility, it also serves clients that are not located in the facility through virtual incubation or incubation without walls.

The Centre regularly conducts industry interface and technology promotional programmes for sensitization of entrepreneurs and to identify interested potential candidates for physical and virtual incubation. The clients at ABI gets the privilege of meeting Scientists, Business Manager and Business Associates directly, to discuss and finalize the strategies to be adopted to take the business forward. It is also the peer-to-peer relationships that develop within the incubator, that ensures the delivery of basic services, such as how to actually incorporate a business; what are the legal issues; how to take intellectual property protection; how to do basic accounting and cash flow; how to do business presentations, etc. Those kinds of skills are what are transmitted as part of the incubation process.

The residency period for direct incubatees is normally for two years, extendable by another year in special cases, depending on the progress of incubation. As the business venture becomes mature enough, the concessions and the facilities provided to the incubatee companies will be gradually withdrawn. Each incubatee of the Unit will have to pay to the Institute a charge for utilization of space, at a rate concessional to the benchmark rate, which is the prevailing market rent realizable. Incubatee mentoring will continue in virtual mode after graduation, on need basis.

Facilities Offered

The Incubation Centre provides shared physical workspace within the Institute premises, management and technical assistance, access to financing and other supporting services to the incubating entrepreneurial firms. Incubation facilities under one roof are:

- Furnished office suites within the premises of CIFT, with shared facilities like secretarial assistance, computing, copying, conferencing, video conferencing, broadband internet and communication services.
- Pilot level production lines
- Culinary facility
- Modern laboratory facilities for product testing and quality control
- Physical and digital libraries

Business Services

The business oriented services offered by ABI includes assistance in complying with business regulations and licensing procedures, financing, information services, marketing, and tailor-made services designed for the various tenant enterprises. Incubator clients can also gain special advantage in terms of tax savings through special regulations for Business Incubators. ABI also offers a wide variety of services, with the help of strong associations throughout the Business Incubation Network.

Suggestions

Increasing adoption of new technologies leads to rise in startups

Today, the adoption line of certain new technologies is practically vertical. Innovations and technologies are rapidly introduced into the market and accepted by society. The time it takes for a new technology to reach mainstream adoption is accelerating exponentially, to the point in the future, where a new technology can have greater than 50 per cent market penetration in just a few years, whereas earlier it would take decades.

This is a very large amount of future growth for technology that portends a very bright future for technology businesses, and in particular, technology startups and startup ecosystems. This can also be seen from another perspective-the valuation of the companies responsible for the disruption. Today, we face a high rate of disruption and technology adoption. Apart from that, the emergence of new technologies is increasing over time. Today, new disruptive proposals appear constantly. By definition, a startup means risk and velocity. By definition, a traditional company hates these two elements. Startups are, thus, the prime drivers of risk and the speed.

Public-Private Partnerships can accelerate the economy growth

India is now one of the fastest growing emerging economies of the world. In a developing country like India with agri-based economy, where food insecurity and poverty needs to be seriously addressed, public private partnerships are a relevant vehicle that can take forward technology and improved agricultural practices for enhancing the yield and productivity of the agricultural sector. Public–private partnership in agricultural R&D is increasingly emerging as an effective means of conducting research in frontline areas of science and technology, commercializing new technologies, and deploying new products for the benefit of small-scale farmers, food-insecure consumers, and other marginalized groups. While public sector research has been particularly relevant to resource-limited areas, the private sector is seen as the major player in the irrigated regions. Through the public-private partnerships, public sector research institutes can leverage the valuable private resources, technical expertise and marketing networks that they otherwise lack. While the technology can be developed in the public sector organizations, the private sector, where the necessary experience, competencies and tools are more readily available; can be associated in commercialization, marketing, and distribution. At ICAR-CIFT, these partnerships are made through the ABI Centre activities and it turned out as a co-operative venture between the public and private sectors, built on the expertise of each partner that best meets the public needs for services or infrastructure through the transfer of technologies and resources.

Business Incubators can ensure the success rate of startup ventures

Entrepreneurship is not an easy road and the more you grow the more challenges you face. From awareness and sustainability of businesses to regulations associated with legality and enforceability, the list of challenges is long. Given these challenges, it can be hard to get started on the entrepreneurial journey without guidance and assistance. Incubators can serve as a great platform to give you that much-needed support and push to build a solid foundation. Similar to an incubation chamber, incubators help the seeds of startup ideas germinate and evolve into the beginnings of future organizations. Incubators tend to be slightly long-term associations while startup founders find their calling, while accelerators focus on fast growth in a relatively short time.

The Government of India's 'Startup India' scheme has the India Aspiration Fund which has the right to disburse Rs. 2,500 crores per year, with a total corpus of Rs. 10,000 crores across four years. This is the capital pile organically available for startups as dispensed under the management of various academic institutions. The incubators set up at these institutions have become playschools/nurseries to nurture India's next generation of unicorns.

It is important to assess the entrepreneurial aspects of incubatees

Like any other enterprising business teams, the startups go through many struggles. All of them do not have the expert knowledge in all the relevant areas. Therefore, incubators must evaluate the management capability of the entrepreneurs and assist in finding the right talent for these companies. Especially, when the entrepreneur is a technologist lacking business skills, it is critical that the incubator assists the innovator in finding managers that have the necessary skills to manage a successful entity and take it to the next level. Two things determine whether a business can get off the ground successfully and sustainably: a validated market opportunity with customers willing to pay for an innovative product or service or process; and a product or service or process that addresses such an opportunity. The incubators that can mentor and guide potential startups on the above parameters can ensure that the startups are not aborted or stunted in their growth. As the startup crosses certain milestones, they need subsequent funding to scale up. The business incubators have a systematic action plan to ensure graduation of at least 80 per cent of the incubatees to a promising stage that can attract Venture capital funding and holistic venture success.

Conclusion

India is primed to create socio-economic impact across business systems around the world in the near future. Startups form the core of this movement and need incubators that are set up with a motive to make them into scalable and profitable enterprises. Aggregating this pool of talent from young entrepreneurs in colleges and universities, can be an important tool to harness growth and progress. With the assistance of technology-based incubators, startups are becoming machines of rapid and tremendous growth that provide solution mechanisms. They are manifesting advances with widespread technological innovation at a breakneck pace. With the active participation and contribution of the Government of India, the country's entrepreneurial collective can expect to receive the incubation and support, it needs to truly transform the future of the country's economy.

❑❑❑

Chapter 2

Technology Profiling and Support for Incubators and Startups

C. Anandharamakrishnan

Director, IIFPT, Thanjavur-613005, E-mail: director@iifpt.edu.in

Growing Opportunities for Incubators and Startups

Agriculture is the most important sector of Indian economy. 54.6% of the population is engaged in agriculture and allied activities (Census, 2011) and it accounts for the 17.4% of the country's Gross Value Added (current price 2014-15, 2011-12 series). With the importance of agriculture sector, Government of India is taking several steps for its sustainable development. Steps have been taken to improve soil fertility on a sustainable basis through the soil health card scheme, to provide improved access to irrigation and enhanced water efficiency through Pradhanmantri Gram Sinchai Yojana, to support organic farming through Paramparagat Krishi Vikas Yojana (PKVY) and to support for creation of a unified national agriculture market to boost the incomes of farmers. Government has also initiated the programs like Skill India, Start Up India, Stand Up India, Micro Units Development and Refinance Agency (MUDRA) to Agri-clinics and Agri-business Centres (ACABC) scheme and Udaan to help budding entrepreneurs. Powered by the government policies and strong engagement of the industry and institutions in the last two decades, agriculture is rapidly evolving into agri-business.

In an aim to double farmer's income, India is increasingly looking for the ways to strengthen agricultural production, food processing and marketing avenues through the integration of innovations and latest technologies; thus creating a huge scope for agri-tech and food startups in the country. India has developed a vibrant entrepreneurial landscape to provide enormous support for the startups and business incubators. Due to the lack of technical know-how and lack of availability of technology, the producers and traders sell the raw produce at low prices or trample in the field. This not only causes great economic loss for the farmers but also reduces the availability of food for consumption. India loses 30 to 40% of the fruits and vegetables, 10 to 25% of the durables, semi perishables and products like milk, meat, fish, etc. in the post-harvest chain [1]. Value addition of the raw materials and prevention of the wastages will help to enhance the income and also promote the livelihood of the producers, traders and the country as a whole.

To encourage the incubator and startups in the food processing business and value addition, they have to be technically trained by providing hands-on experiences. They need to work for themselves on processing and value addition of foods. Furthermore, they have to be provided with incubation facilities for processing and market testing their products. To fulfill all these basic needs, IIFPT established a hi-tech food processing incubation-cum-training Centre. IIFPT has thousands of beneficiaries through the skill development training and technical guidelines and successfully generated 84 entrepreneurs throughout India. The spring up from the enterprising young and innovative Startups in the food value chain will create a transformation in the Indian economy because India's economic future lies in encouraging the startups which will bring dynamism, new thinking and create jobs.

About MoFPI

Ministry of Food Processing Industries (MoFPI) is a ministry of Government of India is currently headed by Smt.Harsimrat Kaur Badal, the Union cabinet minister. The ministry was set up in the year 1988 with an aim of developing a strong and vibrant food processing industry thereby increasing the employment opportunity in the country. With the aim of making food processing a national initiative, the MoFPI has a clear goal of attaining the objectives by facilitating and acting as a catalyst to attract quality investments from within India and abroad into this sector. With this overall objective, the Ministry aims to:

- Enhance farmer's income by better utilization and value addition of agricultural produce.
- Minimize wastage at all stages in the food processing chain by the development of infrastructure for storage, transportation and processing of agro-food produce.

- Introduce modern technology into the food processing industries from both, domestic and external sources.
- Encourage R&D in food processing for product & process development and improved packaging.
- Provide policy support, and support for creation of Infrastructure, capacity expansion/upgradation and other supportive measures form the growth of this sectors.
- Promote export of processed food products.

Research and Development institutions functioning under MoFPI are National Institute of Food Technology, Entrepreneurship and Management (NIFTEM) and Indian Institute of Food Processing Technology (IIFPT). MoFPI offers a scheme called Pradhanmantri Kisan Sampada Yojana (Scheme for Agro-Marine Processing and Development of Agro-Processing Clusters) with an allocation of ₹ 6,000 crores for the period 2016-20. The Scheme is a comprehensive package, which will result in creation of modern infrastructure with efficient supply chain management from farm gate to retail outlet. It will not only provide a big boost to the growth of food processing sector in the country but also help in providing better returns to farmers and is a big step towards doubling of farmers income, creating huge employment opportunities, especially in the rural areas, reducing wastage of agricultural produce, increasing the processing level and enhancing the export of the processed foods.

IIFPT: A platform for budding entrepreneurs

Indian Institute of Food Processing Technology (IIFPT) is a pioneer Research & Development and Educational Institution under the aegis of the Ministry of Food Processing Industries, Government of India, located at Thanjavur, Tamil Nadu. The Institute is in existence for 50 years and was formerly known as Paddy Processing Research Center (PPRC). In 2008, the Institute was renamed as Indian Institute of Crop Processing Technology and was upgraded as a National Level Institute and later in March 2017, was re-christened as Indian Institute of Food Processing Technology. The IIFPT diversify its prospects for intensive research and development activities in the areas of agriculture, horticulture, fish, meat, poultry and dairy processing. It takes up challenging issues on food packaging and testing services, nano-technology, cold chain and logistics, computational modelling of food processing systems, 3-D printing of foods, fusion foods, designer foods and non-thermal food processing technologies. The institute is working on 24 on-going externally funded projects which include 18 National, 2 International projects and 4 SRP projects. IIFPT successfully completed its flagship mission programs on onion, which aimed at doubling the income of onion farmers. Post-harvest processing technologies, onion processing equipment and value added products from onion have been developed in this mission. Apart from this, a common food processing incubation center has been initiated to promote entrepreneurship among the onion farming community at Perambalur district, Tamil Nadu.

The Institute began offering degree courses sends undergraduate students for short-term exposure and graduate students for long-term research projects in reputed international institutions and advanced laboratories around the world. These training and research exposures help to create leaders in food processing who can take Indian food processing sector to greater heights. IIFPT also sends the undergraduate students to India's villages to learn the firsthand problems faced by producers and the current status of production, storage, handling, and marketing of foods. Graduates of IIFPT have placed in higher studies both, international and national, government sectors and industries.

IIFPT used to conduct various outreach programs, conference, seminar, workshops and expose to demonstrate the innovative technologies in food processing sector. Thousands of stakeholders like farmers, self-help groups and young entrepreneurs are getting benefitted through the above said programs. Because of all these endeavours, IIFPT was awarded with the FICCI-Excellence in Institutional Social Responsibility at 13th FICCI Higher Education summit, Excellence award–in Recognition of contribution in promotion of food security and research at Indian Technology congress at NIMHANS Convention Centre, Best Institute–Women Skill Development at Skilling India Summit by ASSOCHAM, secured 87th rank in the overall category in National Institute Research Framework, 2017 ranking.

Liaison Office of IIFPT

Setting up own food business or developing an existing enterprise is an expensive and complex process. One of the biggest challenges is finding premises suitable for technical guidance and practical trainings. To support farmers, budding entrepreneurs, self-help groups (SHG), students, and industries. IIFPT established liaison offices in the northern region of India for providing technical guidance, skill development training, incubation and consultancy services.

Liaison office of IIFPT at Guwahati is serving in conducting skill development training programme, entrepreneurship development and students internship training in the field of food processing. It became a landmark in the food processing skill promotion in north-eastern part of India. Through this office, IIFPT is reaching the unreached communities and serves for the upliftment of the people in this region. IIFPT established its another liaison office and Food Processing Business Incubation-cum-Training Centre (FPBITC) in Bathinda (Punjab) to cater to needs of the people by providing training and skill development in the region, where one can transform their food product concepts and ideas into a successful. By these liaison offices, IIFPT expands its horizon and visibility to serve more stakeholders and disseminating the latest technologies to the unreached masses.

Services offered at IIFPT

IIFPT has 12 departments with well-equipped Laboratories Viz. Food Engineering, Food Packaging and System Development, Food Product Development, Food Safety and Quality Testing, Food Biotechnology, Primary Processing Storage and Handling, Computational Modelling and Nano Scale Processing Unit, Technology Dissemination, Food Processing Business Incubation Centre, Workshop and Fabrication Unit, Academics and Human Resource Development, Planning and Monitoring Cell, Industry Academia Cell.

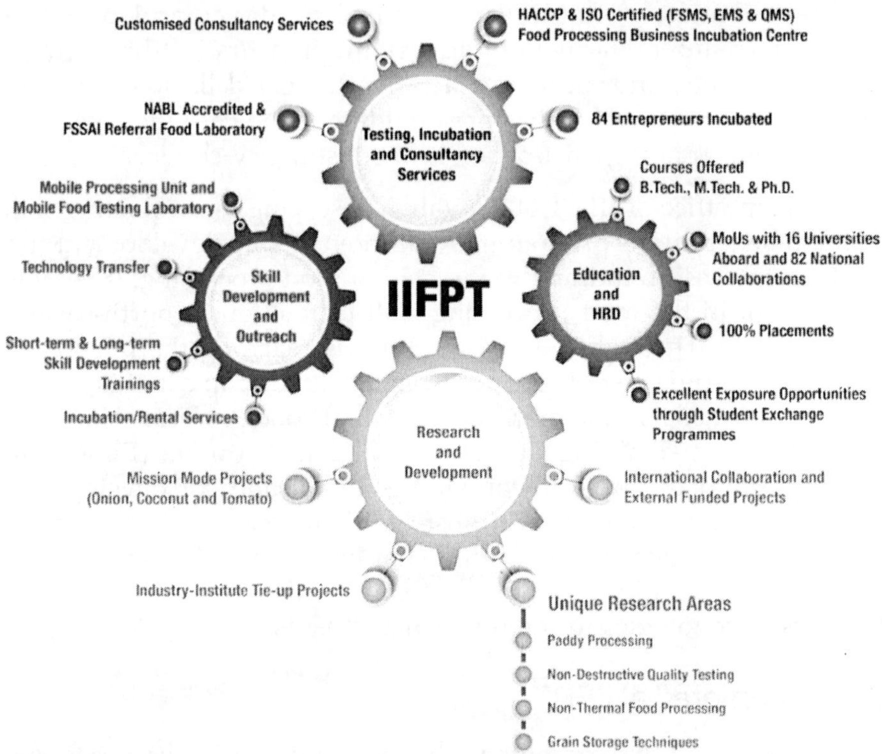

Fig. 1: Services offered at IIFPT

The institute offers academic courses of B. Tech., M. Tech., and Ph.D. in Food Processing Engineering and M. Tech. in Food Science & Technology areas. The institute also delivers food testing services and training through NABL accredited food safety and quality testing laboratory, which also serves as an FSSAI referral laboratory. Incubation and training services are provided through ISO-9000/14000/22000 certified Food Processing Business Incubation-cum-training Centre (FPBIC) and various other extension and consultancy schemes.

Making a processing and testing unit available on or near the farm will encourage the producers to process and preserve the produce for higher economic return at a later date. Setting up a permanent processing and testing unit is nearly impossible due to non-availability of continuous power supply, lack of repair and service facilities near the farm. With this background and a motive to support producers in processing and testing IIFPT, created Mobile Processing Unit (MPU) and Mobile Food Testing Laboratory. IIFPT's mobile processing unit and mobile food testing laboratory aim to provide on-site processing and quality testing services. Mobile Processing Unit can be taken to the farmer's field and process the fruits and vegetables, especially tomato, onion, etc. at the time of surplus production and convert them to intermediate/ final value added products to minimize the losses/ wastage. Similarly, the mobile testing unit can be used for testing the food products of industries those who don't have the in-house testing lab for quality checking of the samples. Details of the above said services are available at www.iifpt.edu.in.

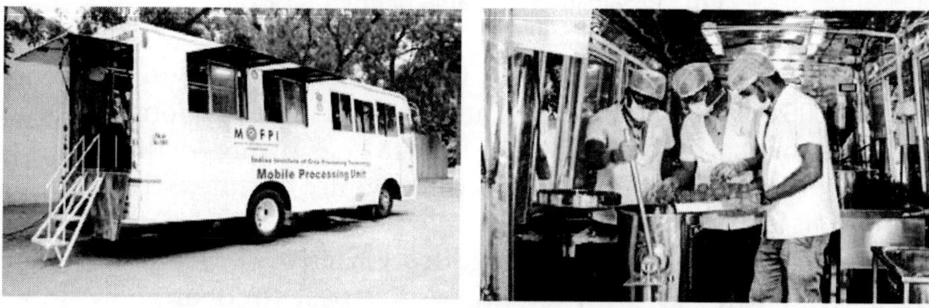

Fig. 2 : IIFPT's Mobile Processing Unit (MPU)

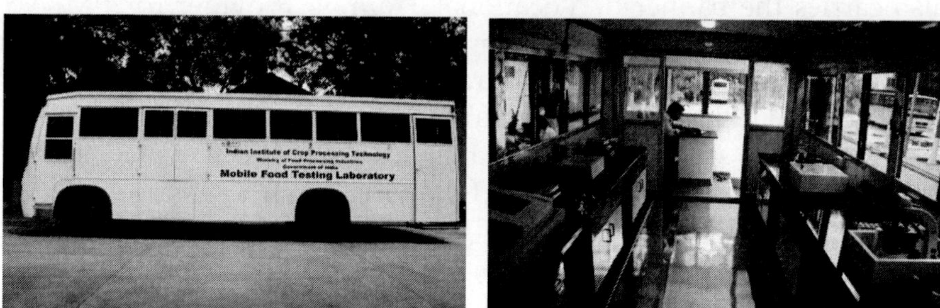

Fig. 3 : IIFPT's Mobile Food Testing Laboratory

Food Processing Business Incubation Centre

The Food Processing Business Incubation Centre (FPBIC) at IIFPT is spread of 7500 Sq.Ft. with the facilities for processing and value addition of agro and horticultural cropsincludes fruits & vegetables, grain and gram, spice processing, cold and hot extrusion line, bakery line with different packaging facilities. Details of the facilities available at FPBIC are given below:

Grain Processing Line

Packaging Line

Fruits & Vegetable Processing Line

Bakery Unit

Spices and Plantation Crop Processing Line

Storage Line

Hot & Cold Extrusion Line

Drying and Dehydration Line

Fig. 3: Facilities available at IIFPT's FPBIC

IIFPT, today, prides itself as the first Indian institute to secure ISO 22000:2005, ISO 14001:2004 and HACCP certifications for Food Business Incubation Centre. ASSOCHAM has awarded IIFPT as the Best Institute Innovation on Skill development at the Skilling India Summit,m 2018 and Best Institute Women Skill Development at the Skilling India Summit, 2017. Integrated Chambers of Commerce and Industry (ICCI) has awarded IIFPT, the Best Innovative Training Provider, 2018. The National Education Summit, 2017 has awarded IIFPT as the "Excellent Institute for Food Processing in India". IIFPT also carries the banner of Vocational Training Provider for PMKVY Programme to conduct training on Baking Technician. FPBIC offers skill development training programmes and consultancy services (Table 1) for budding entrepreneurs as well as the existing industrialists, farmers, SHGs, students and research scholars. FPBIC was designed to be unique to work upon in the following ways:

- Need-based consultancy and support to unorganized food processing sectors.
- Incubation Service.
- Linking Entrepreneurs with the marketers.
- Providing a platform for exhibiting their innovation during any of the Institute activities.
- On-Farm Mobile Processing Unit.

The uniqueness of the incubation centre lies even in its location, at Thanjavur, being the 'Rice bowl of India'. Though the Institute lies in South India, IIFPT has registered clients from all over India. The success story of our entrepreneurs has been so evident that we currently have clients from all over the nation and even international.

Table 1: Training programmes offered at FPBIC

S.No.	Training Title	Duration
1.	Beginners training	1 day, 3 days
2.	Executive training	5 days
3.	In-plant training	1 month
4.	Consultancy idea transfer	1 month
5.	Technical support and guidance	3 months
6.	New product development	3 months
7.	Pradhan Mantri Kaushal Vikas Yojana (PMKVY) - Baking technician training	1 month
8.	Entrepreneurship Skill Development Programme (ESDP)	1 month

Details of the skill development training and other incubation programmes are available at http://www.iifpt.edu.in/departments/incubation-centre.html.

Success stories

Through the various skill development programme and other services, provided by IIFPT, many entrepreneurs have successfully initiated their business. Entrepreneurs were well-equipped through the enormous support provided by IIFPT. Mr. N. Murugesan (Proprietor of Sri NeeVee) and Mrs. Sangeetha Narayanan (Properitor of Lakshya biscuits) have undergone PMKVY training and launched their bakery production unit recently. Overall, IIFPT has 84 successful entrepreneurs, out of that 24 entrepreneurs have started their business within 3 months. The entrepreneurs have started their own processing unit to produce the innovative products like bakery products, millet products, masala powders and pickles, chocolates, breakfast mixes, RTE and RTC products, nannarisarbath, instant tea, soup mixes, RTS fruit beverages, dried and dehydrated products, etc. The entrepreneurs have established their units in the different states of India like Tamil Nadu, Rajasthan, Maharashtra, Odisha and Telangana.

Technology transfer

The following ready to commercialize technologies, developed through R&D works conducted at IIFPT in the past one year, were transferred to different beneficiaries:

1. The technology *"Neera tapping machine"* was transferred to Annamalai Coconut Producers Society, Vinayaka Coconut Producers Society and Pudukkottai Coconut Farmers Producers Company, Tamil Nadu.
2. The technology *"Mobile Processing Unit (MPU)"* was transferred to Tamil Nadu State Agricultural Marketing Board.
3. Tamil Nadu State Agricultural Marketing Board for the technology for *"Small onion peeling machine for onion growing farmers"*, Allathur, Tamil Nadu.
4. The technology *"Non-dairy millet ice-cream"* was transferred to Boinpally, Hyderabad.
5. The technology *"Moringa leaf powder, dehydrated red and white onion and garlic flakes"* was transferred to Maarci Foods, Chennai.
6. The technology *"Non-dairy millet ice cream"* and *"Non-dairy coconut ice-cream"* was transferred to Lantern Foods, Coimbatore, Tamil Nadu.
7. The technology *"Neera sugar crystals"* was transferred to Annamalai Coconut Producers Company, Annamalai, Tamil Nadu.
8. The technology *"Rice ice-cream, rice health drink, GABA rice and edible cutleries"* was transferred to MDJ - Udhayam Agro Foods, Surat, Gujarat.
9. The technology *"Banana juice with suspended basil seeds"* was transferred to Tamil Nadu Banana Producers Company, Thottiam, Tamil Nadu.
10. The technology *"Virgin coconut oil"* was transferred to Pollachi Coconut Producers Company, Pollachi, Tamil Nadu.
11. The technology *"Neera crystals"* was transferred to Pudukottai Coconut Farmers Producer Company Ltd. Pudukkottai, Tamil Nadu.

Mission Onion program

Mission onion program is a classic food processing model, that focused on the end-to-end approach covering post-harvest management, processing and marketing to assure appropriate returns to Shallot onion growers/producers. The infrastructure of the unit created under this model has the entire processing line completely automated for small onion processing, which includes onion root and stem cutting machine, onion grader, peeling machine, and curing-cum-storage structure. This creates facilities to develop value added products like vacuum packed fresh peeled onion, dehydrated products (including flakes and powders), pastes, brined onions, pickles and other allied products from onions. Regular farmer will get an average of ₹ 10,000/- per hectare as profit from small onion cultivation. Through this endeavour, the onion growers can earn a 3-fold increase in profit per acre (₹ 1,50,000/acre). In collaboration with MoFPI and with Tamil Nadu state government, IIFPT established India's first shallot onion processing unit-cum-incubation centre at Perambalur, Tamil Nadu. Common Food Processing Incubation Centre with fully mechanized set up has the facility for processing and value addition of small onion for developing various value added products, such as onion paste, onion powder, onion flakes and vacuum packed peeled onion to extend the shelf life and better price. Around 1000 farmers from the village was joined together to form the Farmers Producing Organization (FPO) and they are only running the common incubation facility. FICCI has awarded IIFPT as the Excellence in Institutional Social Responsibility at the 13th FICCI higher education summit for the mission onion program.

Fig. 4: Automated Shallot onion processing unit

Fig. 5: Shallot onion processing unit-cum-incubation Centre at Perambalur

Mission Coconut program

IIFPT initiated mission coconut programme in collaboration with MoFPI, Micro Small and Medium Enterprises (MSME) and Tamil Nadu state government. The aim of the mission is to develop end-to-end solutions for coconut farmers from harvesting to value addition and thereby to enhance the farmer's income. Presently, a coconut farmer is getting ₹ 1500/tree/year. IIFPT aims to enhance the income to ₹ 15000/tree/year. The institute is planning to develop the entire processing line starting from grading, husk removal, pairing, testa removal, water collection, cutting and slicing of the meat along with value added products from coconut like neera with extended shelf life, vinegar, IQF fresh kernel and vitamins and mineral enriched nata-da-coco, vitamin B-12 fortified copra, virgin coconut oil blend as cooking oil, coconut chips infused with vitamins, coconut water powder, enriched coconut milk, etc. This mission also aims to create an effective supply and value chain by integrating coconut-based common value addition for the use of small and marginal farmers producers companies. As a part of this mission, IIFPT developed the neera tapping machine and increased the shelf life of Neera. In addition to that, IIFPT developed the innovative products like neera crystals and coconut chips.

Contact details

The Director
Indian Institute of Food Processing Technology
Ministry of Food Processing Industries, Government of India
Pudukkottai Road, Thanjavur-613 005
Tamil Nadu
India.
Contact No.: +91 4362 228155
Fax: +91 4362 227971
For general queries on training, testing, consultancy, academics and other services offered by the Institute: info@iifpt.edu.in
For e-mails to the Director: director@iifpt.edu.in
Website:www.iifpt.edu.in

Reference

T. Report, N.A.D. View, R. Hegazy, Post-harvest Situation and Losses in India, (2016). doi:10.6084/m9.figshare.3206851.v1.

❑❑❑

Chapter 3

Customer-Centric "Lean-Agri Startups" for Smart Farming

V.G. Dhanakumar
Director, IIPM, Bengaluru, E-mail: directoriipm@vsnl.net

ABSTRACT

Smart farming is an innovative way of carrying out agricultural activities by reducing human efforts and making maximum utilization of existing resources, with automation system and sensing devices. It is the unique way to increase the quality and quantity of smart farming system through Disruptive Innovations (DI). DI is an innovation that creates & eventually disrupts established/new market and create a value network to enhance the business to delight the customer. Such profound changes in smart farming may not only emerge with opportunity but also induce big challenges. It is crucial to point them out at an early stage of this revolution to avoid lock-ins. Smart farming-led strategic approaches and its policy should provide a concerted path out of locked-in technologies and practices characterised by strong polarization of lean management, design thinking, open innovation, disruptive start-ups and market segmentation.

In the context of emerging lean start-ups, secondary and corporate smart farming business, where great ideas are ideas that solve a problem in a unique way, usually happen when two ideas meet together for the first time for Dialogue, Discussion and Discourse (3D). Therefore, design thinking becomes effective because it promotes such collaborative approach to bring people from different parts of the agri-business organizations together for their different knowledge, expertise, and input on how we are going to solve a problem. By combining those ingredients with open system approach, we can come up with new and creative ways to solve a problem related to on and off smart farming and institutionalizing "Lean-Agri Startups" towards market orientation.

Discussion here illustrate the process of lean-agri-start up with examples and explain the connectedness of smart faming vision with disruptive lean start-up business model using a component of 9 building blocks to form a smart farming business startups through hypothetical approaches, as decision support system.

Keywords: Disruptive Smart Farming, Locked-in-Technologies, Lean Start-up, Design Thinking, Open Innovation, Business Model Canvas (BMC)

Smart Farming is the way to increase the quality and quantity of agricultural production by using sensing technology to make farms more "intelligent" and more connected is called 'precision agriculture' also known as 'smart farming'. Smart Farming is a farming management concept using modern technology to increase the quantity and quality of agricultural products. Farmers, in the 21st century, have access to GPS, soil scanning, data management, and Internet of Things technologies. By precisely measuring variations within a field and adapting the strategy accordingly, farmers can greatly increase the effectiveness of pesticides and fertilizers, and use them more selectively. Similarly, using Smart Farming techniques, farmers can better monitor the needs of individual animals and adjust their nutrition correspondingly, thereby preventing disease and enhancing herd health.

Smart farming can provide a concerted path out of locked-in technologies and practices characterized by strong polarization and market segmentation. It offers a path towards sustainable smart farming by diversification of technologies, crop and livestock production systems, and networks across all actors of the agri-food sector. There is no single policy approach that can achieve this vision, which supports and facilitates the appropriate use of ICT technology. Rather, the idea is to identify the dominant mechanisms that constrain or threaten a sustainable use of the technology and to select the most appropriate actions. This may result in better access to capital in some cases and to specific support of investments in others. Moreover, the support of co-operatively used farm monitoring technology (e.g., jointly owned unmanned aerial vehicles monitoring fields of entire villages) or investments in education and training may also support the sustainable use of these technologies. In all cases, however, the policy environment should provide a clear, legal setting that allows for effective ownership and user rights.

Design Thinking

Design Thinking is a methodology that provides a solution based approach to solving problems of smart farming and welcomes the multi-dimensionality of a dynamic process and one has the liberty to juggle the stages in a manner that complies their workflow or turn the thinking process. The process brings out more ideas and refers more choices within smart farming system. The design thinking process is more of a diverging and converging process as against the linear innovation process. Design thinking methodology is used to capture the user's (smart farming customers) requirements through disruptive innovation(s), as depicted in the Figure 1.

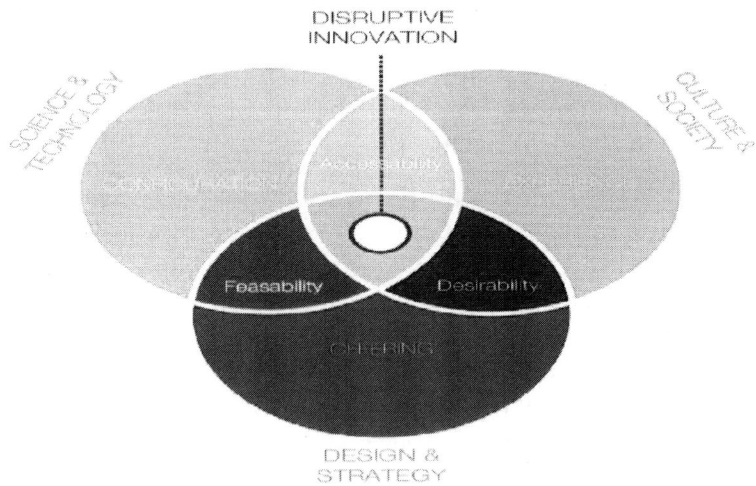

Fig. 1 : Model of Design Thinking
Source: The Many Different Flavours of Design Thinking, Factotum (2018)

Lean Startup

Both design thinking and smart farming process leads to Lean Startup movement, which has changed the global startup landscape over the last few years, providing a simple methodology for resource-constrained smart farming business units to quickly learn from their customers, iterate their product development, and ultimately create products that people want. As a result, lean startup principles are a key component of the smart agri-business vision 2050. According to Adeo Ressi (2016), once Lean Startup Company established its vision, that vision becomes greater than an individual. It is now an enterprise and it should guide each and every decision of the startup makes. An individual vision is the embodiment of the brand, and the reason startup exists. In nutshell, good smart farming business leaders create a vision, articulate the vision, passionately own the vision, and relentlessly drive it to completion."

Successful smart farming startup business models depend on developing three "best strategies," or qualities that help the business succeed through: finding *high-value customers, offering significant value to customers,* and *delivering significant margins.* Best business models also avoid three "bad approaches" that can derail a business: *difficulties in satisfying customers, trouble maintaining market position,* and *problems generating funding for growth.* This highlights the outlines of key factors in determining whether startup business model meets each best strategies and avoids the approaches. Market-based activities are now recognised by governments, business and development agencies around the world as potential solutions to major sustainable development challenges. In fact, the field of strategy has evolved substantially in the past twenty-five years.

The lean startup takes an innovative look at entrepreneurship within startups as well as established business approach at creating business development cycles. The term "lean startup" is derived from principles of lean manufacturing, a manufacturing philosophy mainly originated from the Toyota Production System (TPS) that is centered on the aim of identification and minimization of waste (Emiliani, 2006). Waste is defined as 'any human activity, which absorbs resources but creates no value' (Womack & Jones, 2003, pp.15). In the context of a smart farming startup, waste is described as anything that inhibits the team from learning about how to create value for customers (Ries, 2011). The term "customers" includes all the external actors (e.g. individuals, companies and organizations) for which the startups solution potentially could be applicable. The approach does also draw upon principles of other management theories, such as agile development, design thinking and lean product development as a triangle approach.

Open Innovation (OI)

The process of design thinking, lean startup, agile development, facilitate the process of open innovation, as a paradigm that assumes that smart farming units can and should source innovation from external actors as well as their own internal R&D departments or field-oriented innovations. Alternatively, it is "innovating with smart farming partners by sharing risk and reward." The boundaries between a smart farming firm and its environment have become more permeable; innovations can easily transfer inward and outward. In simple words, engaging in Open Innovation means that the smart farming unit is reaching out far beyond its own internal boundaries in order to innovate, involving external players, working with customers, suppliers, partners, startups, universities, academic research, private institutions and other players in the ecosystem. This enables the smart farming unit to harness external ideas, gain access to bright minds and ideas outside its own enterprise, from different parts of the world, and as a result, dramatically increase its own ability to innovate as illustrated in the Figure 2.

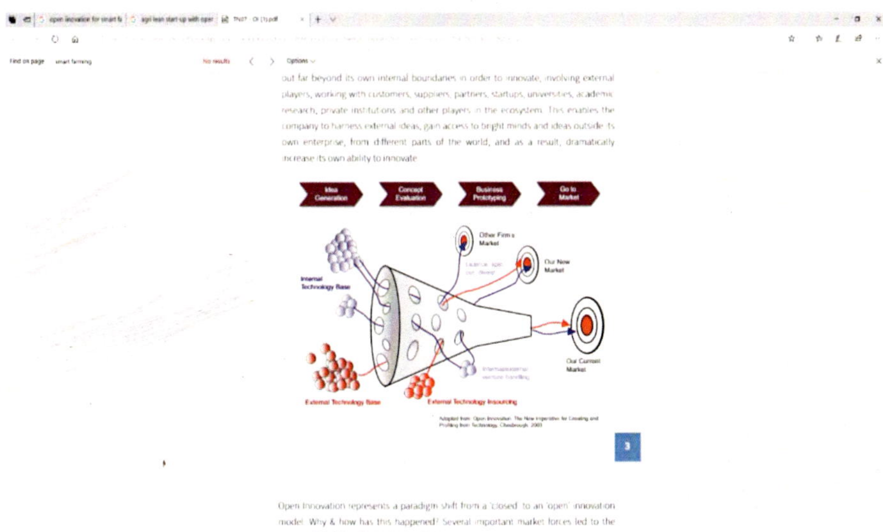

Fig. 2 : Open Innovation
Source: Open Innovation: A Practical Guide, May 2016

In the above process, the open innovation represents a paradigm shift from a 'closed' (conventional farming) to an 'open' (smart farming) innovation model to several important market forces led to the erosion of the 'closed' type of innovation. Open innovation (OI) is becoming inevitable to deal with these challenges and for long-term competitive advantage of the smart farming Agri-Food units. Start-ups can play a key role in corporate's open innovation strategies. Although corporate-startup collaborations (CSCs) are already more prevalent in research and business practice in hi-tech sectors, this phenomenon is new in the smart farming Agri-Food sector. By combining literature on CSCs in hi-tech sectors with OI literature in smart farming Agri-Food, a framework was developed to initiate and manage CSCs. Best practices should focus on alliance formation, strategic fit, governance mode, access to resources, relationship & trust, and IP protection. A retrospective case study on Start-Life and Food-case confirmed that a CSC could be key in improving innovation performance and increasing competitive advantage. It also confirmed the importance of mediating and moderating factors influencing how the CSC leads to competitive outcomes, and that a dynamic view on the CSC is required under smart farming.

Disruption in Smart Farming

The above initiative disrupt smart agri-farming that ultimately comes back to data. All of the innovative solutions used on smart farms today are geared towards collecting and then utilising data for the improvement of productivity. This data has helped and will continue to help smart farmers to identify important changes in crops, livestock and general farm processes.

This is even the case for farmers in the developing world, who can access simple but useful apps for weather and market prices. Future technology will undoubtedly cause further disruption. Genomics, for instance, is a sub-discipline of genetics, which has received mass funding in the last few years. Genomics analyses and then creates genetic data, which could theoretically develop entirely new plant and animal species. As more things become connected by Internet of Things (IoT), it will be easier to track and predict events on the farm. Robotics, increasingly equipped with artificial intelligence (AI), could streamline automated processes even more. This will inadvertently lead to farms that don't rely on human labour, thereby changing traditional employment patterns. Aided by advanced robotics and autonomous systems, farming will become less of a manual job. Instead, smart farming personnel will focus on engineering, programming and mechanical skills. This disruption will be disconcerting for traditional farmers, but their productivity is likely to suffer, if they are slow on the uptake. As well as providing food for the world, farms are businesses which need to adapt to survive.

Agriculture has undergone continual disruption, evolving from an entirely manual industry into a largely mechanical one. Smart Farming, fuelled by innovative technology, will only accelerate this trend. Data collection, advanced robotics, machine learning and genomics are just a handful of the technologies primed to maximise the productive potential of today's farms. In fact, thanks to connective technology, farmers won't even need to be on the farm to know what's happening within it. Accelerating the adoption of tech within farming is ultimately beneficial to everyone, as without agriculture, there would be no food on our plates. Unfortunately, disrupting legacy industries is facing numbers difficult due to outdated infrastructures and conservative opinions. As population numbers creep steadily, hopefully this will change. To sum-up, a total process of design thinking, lean start-up, agile development, open innovation and disruptive business leads to formation of smart farming business canvas model for designing agile development product.

Business Module Canvas (BMC) Technique: Hypothetical Sketch out for new product

In smart farming establishment of learn startup units are currently very limited. Most smart farming in the startup ecosystem today, used on how they leverage within startup ecosystem. In most cases, such integration is to make smart farming based products to be special and exciting. However, the fact is most smart farming units do not have the resources or the talent needed to leverage smart farming in a way that would add to their value proposition to the sector. This section is to help smart farming beneficiaries to understand whether their particular business model needs smart farming from the start, or it is something that may evolve in near future, as it is not required for the basic value proposition.

A good hypothesis is important because it leads to good experimental design. Good experimental design is important because it needs to properly validate or invalidate what we are doing is the process of smart farming business. In the past, business initiator plan business plan a static document that describes the size of an opportunity, the issue to be solved, and the solution that new venture will provide. It includes a 3-5 years forecast for income, profits and cashflow. Typically, a business plan is essentially a research exercise written in isolation before an entrepreneur/startup has even begun to build a product. Now, after decades of examining startups process by the entrepreneurs agree that instead of writing an intricate business plan, they should initiate hypothesis in a framework called "**Business Model Canvas (BMC)**" along with its 9 building blocks as indicated in the Table 1 under Annexe-1. Under the BMC, it contains the component of smart farming business production through hypothesis to be tested to understand how a business startups create value for itself and its customers related to smart farming enterprises viz., spices, tea, coffee, coconut & dairy business domain and use of agile (quickly/ easily designing products in a phased manner) development to minimise total number of days in the planning and create a Minimum Viable Product (MVP) for testing and launch Minimal Marketable Product (MMP) within smart farming ecosystem. During customer development process, a smart farming start-ups unit searches for a right business model hypothesis, based on customer feedback. This process either revises the plan or "pivots" to new hypothesis until it finds a right approach.

Table 1: Business Model Canvas (BMC) consists of all Nine Building Blocks of Agri-Smart-Farming-BusinessProcess

Key Partners	Key Activities'	Value Prepositions	Customer Relationships	Customer Segments
Who are our key partners? Who are our key suppliers? Which key resources are we acquiring from our partners? Which key activities do partners perform?	What key activities do our value propositions require? Our distribution channels? Customers' relationships? Revenue streams	What value do we delivery to the customer? Which one of our customers' problems are we helping to solve? What bundles of products and services are we offering to each segment? Which customer needs are we satisfying? What is the minimum viable product?	How do we get, keep, and grow customer? Which customer relationships have we established? How are they integrated with the rest of our business model? How costly are they?	From whom are we creating value? Who are our most important customer? What are the customer archetype?
	Key Resources		**Channels**	
	What key resources do our value propositions require? Our distribution channels? Customer relationships? Revenue streams?		Through which channels do our customer segments want to be reached? How do other companies reach them now? Which ones work best? Which ones are most cost-efficient? How are we integrating them with customer routines?	
Cost Structure			**Revenue Streams**	
What are the most important costs inherent to our business model? Which key resources are most expensive? Which key activities are most expensive?			For what value are our customers really willing to pay? For what do they currently pay? What is the revenue model? What are the pricing tactics?	

Source: Modified version of Why the Lean Start-up Changes Everything (Steve Blank), 2013

Table 2 & 3 of Annexe-1, identifies and assertions about the actions that should be carried out as part of Business Model Canvas (BMC) related to spices/tea and diary sector using a components of 9 building blocks to load smart farming business startups through hypothetical approaches.

Table 2: Business Model Canvas (BMC) for Spices/Tea Sector Field and Retail Operations (FRO) Organic Product (Spices/Tea)

Key Partners	**Key Activities'**	**Value Prepositions**	**Customer Relationships**	**Customer Segments**
Certified organic producers & retailers	Branding	Products free form biological, chemical, physical & allergents (safe products)	Trust building for organic product	Household
Wholeseller	Assortment	Affordable price		Restaurant
	B2C distribution	Research innovations	Co-creation of community & retails	Hospital
Stores				Anganvaadi
Web sales	R&D innovation			
FPOs	**Key Resources**		**Channels**	
Organic Family Gardens (OFGs)		Retail and Organic Certification (ROC)		Organic
		Packaging	Organic retail outlet	product
	Patent			promotion
	Marketing		Direct sales	
	Health professionals		Media	
			Women consumer channel	

Cost Structure	Revenue Streams
Lean Management	Optimum profit & repetitive sales
Clever pricing	Start-up for organic product

Source: Modified version of PPT Presentation of Selected Business Model Case Studies of Rural Service Providers in Bangladesh.

Table 3: Business Model Canvas (BMC) toward a Multi-Market Approach (Dairy Sector)

Key Partners	Key Activities	Value Proposition	Customer Relationships	Customer Segments
Primary societies Marketing division Distribution agents	Milk collection, processing shipping Veterinary services Selling feed at no profit	Convenience & brand power Value added products: ghee, butter, powdered milk, curds Risk reduction/security: MV will always buy, where as other companies will leave the market when price drop Accessibility: cow loans	Self-service for dairy consumers Personal veterinary assistance Community/co-creation: profit sharing & targets	Dairy product consumers Coop members Families of employees (School & other benefits on company site)
	Key Resources		**Channels**	
	Dairy shed (30 km radius) Factory equipment Transportation network		Factory design (Collection & processing) Society division (veterinary corps) Advertisements	
Cost Structure		**Revenue Streams**		
Pay for milk, based on fat content Milk transport price determined by amount' 70% government owned Veterinary service (10+ in the field every day) facing a shortage		Income from selling processed goods in urban and rural markets Sold by distribution agents who earn commission but have to sell at the price set by government 5% service fee on cow loans		

Source: Modified version of PPT Presentation of Selected Business Model Case Studies of Rural Service Providers in Bangladesh.

But, learn smart farming startup methodology is not a simple process. It is a hypothesis driven approach to learning the most important things about the business startup. In a nutshell, a hypothesis is our proposed belief, based on limited evidence to make some business decision. The increased focus on business models could be related to the large number of product and service innovations displaying disruptive characteristics, but where these disruptive features stem from different ways of performing business activities rather than from technological characteristics in the process of startups. The term Disruptive Innovation (DI) model describe for those who are unfamiliar with new innovation/product. Christensen, M. *et al.* (2003) termed these sustaining innovations, because they sustain established performance improvement trajectories by offering demanding customers better performance. Under DI new entrance, for example spice-led startup to the market would win, when an innovation involved making a new spice product or process that current customers actually could not use because it seem to have two little performance, even though it was smaller and more flexible. A new startup in the spice sector would seek out non-traditional markets that uniquely valued the innovation, despite the limitation that made it unattractive to the main stream spices consumer. This process terms as disruptive innovations. In some, incumbents tend to win sustaining battle; entrance tended to win disruptive ones.

Conclusion

To conclude that the exercise of designing new business models is closer to an art than to a science. From an academic point of view, we believe that having clear definitions of the constructs that we employ and analyse in business process is a necessary condition for progress in the field. From the practitioners' perspective, it is our hope that having an integrative framework that cleanly separates the realm of strategy, business models, and tactics will help guide the search of novel, interesting, and profitable new ways to compete. As illustrated in the paper related to lean startup for smart farming, the concept of Business Model Canvas (BMC) as the seamless coming together is elaborated with the key components of the smart farming business startup to leverage and its distinctive value proposition for the tea/ spices and dairy sectors. The paper also highlighted the design of creating BMC-led hypothesis for startup with Minimal Marketable Product (MMP) in smart farming within the triangle approaches of design thinking, lean startup and open innovation for disruptive business process.

Education Implication

The smart farming features of future policy tools tend to combine availability of financial support with specialised advisory services on lean startups and management training, as well as infrastructure provision. This allows to offer the smart farming beneficiaries and integrated support towards sustainability.

In addition, HR and R&D organizations as well as smart farming startups units need to inculcate "The Lean Culture" for a successful lean implementation viz., Lean Concepts, Lean Culture, Lean Planning, and Lean Tools for Lean startup activity across the smart farming business ecosystem framework. The main policy approaches to address challenges associated to smart farming, lean startups policy intervention should deals with:

- Policy rationale and objectives for hypothetical process of smart farming business interventions
- Promoting lean smart farming startup as the welfare state is the fabric of smart farming polices
- Ensuring coherence of policy interventions on "designing lean-smart-farming and startup-entrepreneurs" as a viability of smart entrepreneurship
- HR interventions on 5C (Deficit of "Managerial Capital" within business viz., Cash, Customers, Competition, Capital and Consultants aspects).

References

Achim Walter, *et al.*, 2017. Smart farming is key to developing sustainable agriculture. PNAS. 6148-6150. Vol. 114. No. 24.

Adeo Ressi., 2016. Create A Clear Vision for Your Business, and then Stick to it. Forbes.

Ashiq MI. *et al.*, 2017. Implement design thinking methodology for prototyping a smart farm application to monitor, control and view real-time data analytics through SAP BUILD. International Journal of Research and Scientific Innovations (IJRSI). Vol IV. Issue XII. 2321-2705.

Christensen, M. *et al.*, 2003. The Innovators Solution. Boston; Harvard Business School.

Dhanakumar, V.G., 2017. An Analysis of Disruptive and Sustainable Technology (DST) in India's Plantation Sector. UEM CONGRESS. Bangkok.

Dhanakumar, V.G., 2018. Creating a Disruptive Agri-Business Model for Sustainability. National Seminar on Promotion of Dairy Entrepreneurship and Skill Development for Doubling Farmer's Income. National Dairy Research India (NDRI). Bengaluru.

Dhanakumar, V.G., 2018. Lean Startup: An Agri-business Perspective.

Emiliani, M.L., 2006. Origins of Lean management in America-The Role of Connecticut businesses. Journal of Management History. Vol.12. Issue 2, pp. 167-184.

Ries, E., 2011. The Lean Startup: How Constant Innovation Creates Radically Successful Businesses. Penguin Group, London.

Steve Blank, 2013. Why the Lean Startup Changes Everything. Harvard Business Review.

Womack, J.P. and Jones, D.T., 2003. Lean thinking–Banish Waste and Create Wealth in Your Corporation. 2nd ed. New York: Free Press 1.

❑❑❑

Chapter **4**

Strategies for Agri Startup Initiative for Small Farmers

S. Prabhu Kumar

Former Zonal Project Director, ICAR- ZPD, Ludhiana, Punjab, India

E-mail: prabhukumar@gmail.com

After the successful launch of the milk co-operatives during 1960s, which later brought in the White Revolution, we sowed the seeds of the Green Revolution. While the government imported high-yielding seeds of dwarf wheat from Mexico, and made available irrigation along with external inputs like chemical fertiliser and pesticides, farmers did the rest. In 1967, the first harvest after the Green Revolution technology was introduced was a record three million tonnes higher. Since then the country has not looked back. From an era of food imports, India graduated to food self-sufficiency. The 'glorious' period for farmers lasted for a decade-and-a-half. Although the Green Revolution had by-passed small farmers, an effort was made to paint a rosy picture of prosperity. The image of a progressive farmer driving a tractor was flashed as a sign of prosperity. In reality, the increase in production did not commensurate with an accompanying increase in farm incomes. While the successive governments were content with bumper harvests, farming as a community remained neglected. Coupled with a declining rate of public sector investments, the demise of agriculture began soon after the mid-1980s. The dismal trend has since continued. While farmers were denied their rightful income, huge salary jumps were provided to other sections of the society. From a monthly salary of Rs 90 per month in 1970, the salary of school teachers, for instance, jumped by 280-320 times by the year 2015, during a period of 45 years. In the same period, salary of government employees went up by 120-150 times; and that of college professors by 150-170 times. Wheat price for farmers on the other hand increased by a paltry 19 times in the same period. Agriculture turned uneconomical, and repeated demands for providing a level playing field fell on deaf ears. For nearly three decades, more so after economic reforms were ushered in, agriculture has been a victim of deliberate neglect and apathy. Successive governments had deliberately created conditions turning farming non-viable thereby forcing an increasing number of farmers to abandon agriculture and migrate to cities. Agriculture is being killed deliberately to keep economic reforms going. To achieve economic growth, mainline economists tell us that it is absolutely essential to move bulk of the population from agriculture to the cities. Food can be produced by promoting corporate farming or can be imported.

Present status and needed approach

Status of present farming include smaller acreage farms, compared to national average, increasing land prices per acre, increasing production costs, growing disconnect between value of farm production and retail food prices. Under this situation, we should concentrate on the strategies for promoting **Agri Startups** among the farming community.

Indian agriculture is **Complex, Diverse and Risk Prone**. The agricultural development activities should address all these three issues for sustainable livelihood security for the farmers. Under this situation, we should concentrate especially on better prices, cost reduction, mitigating risk and improve supply chain efficiency. The focus should be on addressing **3Ms** in agriculture i.e., **Market, Monsoon and Mindset.**

Strategies

The need of the hour is changing from supply driven extension to demand driven extension. Mainly the strategies for promoting Agri startups should be more on empowering pride of ownership with profit and prestige to the farming occupation. Under this situation, technocrats in agriculture should think beyond the technology delivery as farmers are very much conversant with technology part and they need more empowerment in taking the farming as a way of business rather than a way of life. In this regard, the following strategies, namely Speciality Agriculture, Smart Agriculture for Smart Farmers, Precision Agriculture, A model for FARMERS' PRIDE and Agro Tourismare suggested for the empowerment.

Speciality Agriculture

In agriculture, development can happen in three phases. They are:

i) Primary Agriculture

ii) Secondary Agriculture

iii) Speciality Agriculture

Primary Agriculture refers to production of agricultural commodities. **Secondary Agriculture** is the process of making money with the agricultural commodities through processing, value addition and marketing. **Speciality Agriculture** is adding value to a product or process.

Speciality Agriculture can be briefly described as non-traditional agribusiness products, including commodity items not traditional, value-added agri-business products or processes and traditional agri-business products which are marketed or distributed using new/ innovative channels.

Examples of Specialty Agriculture include extended season produce, identity preserved grains, organic products, bio-fuels, further processed food items, alternative forest products and agri-tourism/ agritainment.

Importance of Specialty Agriculture covers potential for greater return on investment or greater return per acre than traditional commodity items, allows agri-business producers to take advantage of competitive advantages gained through diversification and/or differentiation and greater long-term growth potential for specialty agriculture products than traditional commodity items.

Specialty Agriculture trends around creating value vs. capturing value. Capturing value is a function of altering the distribution of value in the agricultural production chain, or by cutting costs *(vertical integration, FPOs)*. Creating value is a function of altering the actual or perceived value to a customer *(unique or branded products, identity preserved or specialty products, bundling, such as agri-tourism)*.

In other words **Speciality Agriculture** confines to precision agriculture *(protected environment, assured production, high-quality products, assured market, higher rates)* as compared to traditional *farming (depends on nature, uncertainty of crop, low quality produce, non remunerative prices, lack of markets)*.

Vision of Speciality Agriculture is to ensure at least a minimum income of Rupees One thousand per day by a family from one acre of land. Farmers can achieve this, if they are supported with relevant advice on production and marketing. Farming is a lucrative option and in speciality agriculture educated youth should come forward to farming. Small farmers, who dominate Indian agriculture, need to be organized and the organizational structure of a producer company will help small farmers to realize better prices. Through adoption of hi-tech farming, one can manage land, labour, power and other inputs efficiently.

Smart Agriculture for Smart Farmers

Smart Agriculture is not a new concept but has been evolved as a result of complex problems arising due to factors, which can be mitigated by efforts, analysis of situation (diagnosis based prescription) and programmes that meet such requirement and flexible enough to change as per required situation. The focus has to be small stakeholders and uneducated farmers, women farmers and rural youth :

- Needed support
- Sharing technical knowledge
- Providing credit, subsidy and safety net
- Monitoring and co-ordination of different agencies
- Providing information to farmers
- Available option for crops based on soil health and moisture analysis, for markets - local & district levels
- About climate - weather forecasting - long - medium - short term day to day. This is should be supported by agro - advisory
- Based on forecast
- After unforeseen climate event has taken place like heavy rains and flooding of fields
- About opportunity to have multiple source of income
- Different players in agriculture sector need completely different types of climate advise

The new extension mechanism has to keep the following in view:

a) First is need prior to commencement crop season - what they should grow - pre-kharif (monsoon) and pre-rabi (winter) based on prevailing climate parameters.

b) This has to be followed by intra - weather agro advisory - based on unforeseen changes in weather during season (contingency plan).

c) Finally at post harvest stage - where they sell products or store it till the market price is improved.

d) Farmer also needs to know what is market demand - both as regards to quality and in terms of overall requirement and possible market price. This will enable him to grow right type of crop.

e) Similar advisory is needed for livestock management and fisheries. Hence, entire exercise is multi- dimensional and extension officer have to be trained for that.

Key Recommendations for Smart Agriculture

1. Unpredictable and erratic climatic patterns resulting from climate change will affect crop production. This will have an impact on farmer livelihoods and food availability. Climate-Smart agriculture provides management options to farmers to both adapt to, and mitigate climate change and maintain his income and look for opportunity to increase it.
2. Crop production must adapt - crop varietal selection, plant breeding, cropping patterns and ecosystem management approaches and become resilient to changes (frequency and intensity).
3. Crop production can contribute to mitigating Climate Change by reducing green house gas (GHG) emissions - for example by reducing the use of/ judiciously using inorganic fertilizers, avoiding soil compaction or flooding to reduce methane emissions (e.g. in paddy rice systems) and sequestering carbon (e.g. planting perennial crops and grass species).
4. Farmers are the primary custodians of knowledge about their environment, agro-ecosystems, crops and cropping patterns, and local climatic patterns. Adapting cropping practices and approaches will be related to local farmers' knowledge, requirements and priorities. Sustainable crop production provides

farmers with options for farming sustainably, taking into account the local ecosystem and how to follow selection of crops, which can be sustained by soil-based on soil health and moisture analysis of their land.

5. Integrated approaches-such as crop-livestock systems, rice-fish systems and agroforestry-diversify food sources and consequently strengthen the resilience of farmers' livelihoods. They also provide opportunities for mitigating climate change but more precisely also increase their income level and in worst circumstances maintain it.
6. Agriculture extension should be conceived as model of inter-dependence within specific innovation system framework of the stakeholders and institutional context based on the strengths of both public and private sector.
7. Technology dissemination process should be unified production system approach, knowledge centric approach and research-extension-farmer interface.
8. Capacity building programmes to train farmers for seed production and storage for self-pollinated crops in cereals, millets, pulses and oil seeds.
9. Establishment of Bio Pharmacy in order to provide quality bio products like *trichoderma, pseudomonas*, vermicompost, botanicals and microorganism, etc., for promoting organic agriculture.
10. Developing entrepreneurship and appropriately synthesizing with extension so as to have a meaningful and relevant extension system that suits to the entrepreneurs who are engaged in agriculture, livestock and related activities.
11. Support to Farmers Producer Organizations (FPOs) for building a strong foundation of producer initiative and ownership. They need long-term efforts and provision of hand holding support. The FPOs should be linked to the markets through public-private partnership mode.
12. For an effective dissemination of smart agriculture to smart farmers, there is need for a unified and modified extension system entitled **Integrtaed Farming Extension System (IFES)**, which focusses on periodic training of subject matter specialists and officers of the Development Departments and introduction of multi-purpose village extension worker for each gram panchayat.

Precision Agriculture

Precision Agriculture or Precision Farming (PF) is a concept of using the new technologies and collected field information, doing the right thing, in the right place, at the right time. Collected information may be used to more precisely evaluate optimum sowing density, estimate fertilizers and other inputs needs, and to more accurately predict crop yields. It helps in avoiding unwanted practices to a crop, regardless of local soil/climate conditions, i.e., it reduces labour, water, inputs, such as fertilizers, pesticides, etc. and assures quality produce.

Role of Farmer Associations: The farmers associations appear to perform two vital roles in the dissemination and success of PF technology. The associations serve as nodes for exchanging knowledge and information. They also help farmers obtain better value for produce as well as inputs. The associations perform a vital support function as information nodes. The association holds regular monthly meetings to discuss marketing and other issues. The associations act as demonstration vehicles to disseminate both, knowledge and information about the techniques and the impact that they have on cultivation and post-harvest results. They appear to help reduce the knowledge and information deficit. The associations also seem to help the farmers obtain better value by improving their negotiating position vis-à-vis buyers or input providers. The organized markets increasingly prefer to deal with farmer associations, as it helps to eliminate risks of delivery failure while providing a greater assurance of quality. This is also beneficial to the farmers, as it helps them to secure better value by costing out delivery failures and in-transit damage to produce out of the revenue. By assuring minimum quality through proper grading and sorting, associations help farmers obtain better average prices than comparable produce sold without the association's involvement. The associations also help the farmers to negotiate better price for inputs, such as fertilizers, pesticides, seeds, etc. by guaranteeing minimum quantity, as well as negotiating for or arranging *timely* supply of inputs. The role of the association was vital in facilitating in the marketing of the produce. The association also acted as the distributor or dealer for several agri-product corporations and sold input materials, such as seeds, fertilizers, plant protection materials, and other agriculture inputs. The products sold were of better quality and at cheaper rates.

Beyond Association Farmers: The impact of the PF must be noticeable in terms of the extent to which the PF technologies travelled beyond the beneficiary farmers. The manner, in which the precision farming technologies travelled to the farming community and the impact, they had on the farmers. In terms of the travel of technologies, travelled unmodified from the producing domain to the receiving domain (beneficiary farmers). The extent of travel was complete in the sense that all beneficiary farmers adopted and continue to use this technology.

Under precision farming adding value to social capital is the highlight. We have to allow the farmers to understand the ground realities and involve them in all the decision-making process. To achieve this, we have to adopt farmer-to-farmer strategy, which will result in bargaining power, negotiating power, commercial activity, total quality management, market study, organised visits, capacity development, self-sustainability and develop linkages.

Hence, the output of precision agriculture will be:

- Doubling the productivity of crops
- Around 95 per cent marketable produce with less unmarketable produces
- High end quality parameters and 30 per cent premium price in the market
- Water economy 30 to 40 per cent
- Electricity economy 50 per cent
- Extended period of harvest to match lean season in the market
- Less labour dependence
- Extended shelf life
- Roughly 25 per cent more weight per unit volume for the produce
- Empowerment of farmers - (*Technical, Economical and Social empowerment*).

A model for FARMERS' PRIDE

Majority of farmers rely on agro dealers for access to farm inputs and services, such as seeds, fertilizers, pesticides, veterinary drugs, animal feeds, mineral supplements and farm machinery. However, most agro dealers who occupy this critical space in the agricultural value chain, are limited in their capacity to address the needs and demands of the farmers that depend on them. The limited capacity of agro dealers blocks the potential of success for the vast majority of smallholder farmers who rely on them.

Farmers' Pride idea is to transform agro dealer outlets into professionally run businesses and leverage that to build a sustainable franchise model business. Professionalizing agro dealers also means giving them access to technology, both hardware and software, to allow them to operate more efficiently. Working with Farmers' Pride, each agro dealer will benefit from a wide range of franchise benefits and guaranteed increase in sales and traffic. Essentially, they will be transformed into businesses that banks and manufacturers are comfortable working with.

Farmers' Pride is an innovative solution to smallholder farmers. Farmers' Pride is providing rural smallholder farmers with timely and relevant quality farm inputs, services and information at affordable prices through comprehensive and innovative franchise model of agro dealers. The business model integrates farmer field extension and education services, quality agricultural inputs sales and distribution, farmer market linkages, financial services, soil testing services and farmer insurance services to offer a reinforced platform for rural social and economic development. Farmers' Pride is leveraging on professionalizing agro dealers to build a sustainable social enterprise.

Farmers' Pride builds and manages relationships with leading agricultural inputs suppliers and manufacturers to secure access to quality products and services for agro dealers and smallholder farmers. Farmers pride retains part of negotiated margin.

Farmers' Pride identifies local agro dealer entrepreneurs in rural markets. These partner shops are staffed with trained agronomists and livestock specialists to provide farmers with correct information. Apart from serving as a sales center for products and services, the shops become the center point for a range of extension, training and awareness creation activities including farmer clinics, product promotions and other forms of farmer extension services.

Farmers' Pride is building an ecosystem that enables other products and services use the platform interface as a medium to provide their already created products and services like insurance and soil testing.

Farmers' Pride targets rural smallholder farmers and agro dealer shops. Farmers' Pride typical customer is a mixed farming family. They farm multiple crops throughout the year, and keep some animals. They depend on a variety of inputs and services across different seasons to support their diversity of farming activities. These farmers and agro dealers are empowered through Farmers' Pride ecosystem to become more prosperous and happy.

Agro Tourism

Agri Tourism is to experience the real rural life, taste the local genuine food and get familiar with the various farming tasks. More than a profession or a business, agriculture is India's culture. Hence, adding additional income generating activities to existing agriculture would certainly increase contribution of agriculture in the national GDP. Serious efforts need to be made in this direction and Agri-Tourism is one such activity.

Tourism is termed as an instrument for employment generation, poverty alleviation and sustainable human development. During 1999-2000, direct employment created by tourism was 15.5 million. Besides, tourism also promotes national integration, international understanding and supports local handicrafts and cultural activities. During 2000, the number of foreign tourists that visited India was 26.41 lakh.

What is Agri Tourism?

Agri Tourism is leisurely activities in the rural areas that help a person understand and appreciate the land and the people who live on it. In recent years, travel has become much more than just ticking through the mandatory list of local museums and sites. Travellers want to slow down when they discover a new destination. They want to meet locals in their natural environments and become more involved with the land they are visiting. With this in mind, Agri Tourism was born. At its most basic level, Agri Tourism is a style of travel that takes place on a farm or ranch and usually offers the opportunity to help with on-site farming or ranching tasks during the visit. It entails a meaningful visit with a producer of land-based products and services. Agri Tourism is, however, not all about staying in a village and relishing the food, this is an opportunity to be close to where the 75 per cent of Indians live. One of the best things about staying on a farm is that guests can contribute to the place through their involvement. Our idea is to make tourists live life like a villager, right from milking the cow, ploughing the field, bathing in a well to climbing a tree and plucking fruits.

Why Agri Tourism?

Mother Nature is open door school without brick walls! If observed carefully, one can learn something or the other, moreover India is Agriculturist's Country, it is expected that we should know the information related to agriculture. Today, urban children's world has became limited in the closed door school, classes, cartoon programs on the television, video games, chocolates, soft drinks, spicy fast food, computer, internet, and so on, they see mother nature only on television screen. Now it has become very necessary that children know the traditional way of agricultural farming activities, and other businesses dependant on agriculture. Here children come very close to Mother Nature and learn many new things in life.

Why is Agri Tourism important?

As commercialism and mass production become the standards by which we live, Agri Tourism has given people who work in the agricultural and horticultural sectors a chance to share their work with the masses. Some Agri Tourism experiences allow guests to buy food products grown on the farm or hand-crafted products made by the farmers' families; purchasing these goods helps provide ranchers who rely on their land with another source of income. Home and consumer education has given way to technology courses in middle and high schools, and many children grow up without ever really knowing what the countryside is or what it's like to interact with live farm animals. Agri Tourism, therefore, gives parents the opportunity to introduce their children to something other than the city life.

A Win – Win Situation

Many countries of the world experience decline in growth rate of agriculture including India. Agriculture as a profession has proved to be uneconomical which has resulted in lack of providing pride, profit and prestige to famers. In India, the annual growth rate in agriculture sector is about 3 per cent while that of industry sector is 10 to 12 per cent and of service sector is 30 to 35 per cent.

As a result, the farming community is switching over from agriculture to non-agriculture activities. About 60 per cent of the farmers do not like farming. Moreover, it has also resulted into migration of population from rural to urban areas.

The Speciality Agriculture, Smart Agriculture for Smart Farmers, Precision Agriculture, A model for FARMERS' PRIDE and Agro Tourism can convert these challenges into opportunities through creating a **win-win situation**, reaping benefits to all the concerned stakeholders.

□□□

Chapter 5

Potential of Startups in the Domain of Agricultural Research and Testing

Pitam Chandra

Former Director, ICAR-CIAE, Bhopal, E-mail: pc1952@gmail.com

ABSTRACT

It is an effort to justify the formulation of a policy for promoting the creation of agricultural research and testing laboratories in private domain, in PPP mode or otherwise, to minimize time and money expenditure on such facilities in public academic and research institutions. This would lead to a considerable expansion of research programs with the existing budgetary support because public expenditure on both, permanent and consumable contingencies for research projects would reduce drastically. At the same time, the policy would create opportunities for qualified and skilled human resource to provide the relevant services to researchers, industries, and farmers through service-based startups. Besides, the modified research ecosystem with significantly reduced need for procurement of instruments/equipment would make more scientific man-days available for actual research work, reduce gestation period for achieving results, encourage increased participation of scientific personnel in conducting meaningful research, lead to better utilization of research infrastructure, and ultimately be more sustainable.

Keywords: Research ecosystem, Startups, Policy, Infrastructure, Sustainable research.

Introduction

Agricultural research is becoming expensive day-by-day for the simple reason that expensive instrumentation and equipment are essential to record a large number of relevant observations. As a result, huge sums of money are required to create experimental infrastructure in academic and research institutions for reliable and replicable results. At a time when public funding of agricultural research is not growing, it is becoming increasingly difficult to fulfill the national research agenda. The need to bring private sector on board for partially meeting the agricultural R&D costs has been realized globally (FICCI, 2017; Byerlee and Alex, 1998). Singh (2011) justified the need for private funding in view of increasingly capital intensive nature of agricultural R&D coupled with consistent slowing down in growth of government funding.

When a prospective researcher proposes a research project, there is a component in the proposed budget to procure modern and expensive hardware/software to conduct the research. In fact, the procurement of goods generally constitutes about 20-60% of research funding in any sponsored project. Although the effort required for procurement of goods for the research is immense, sometimes disproportionately high as compared to the actual research effort, the researcher goes through the process because there is no alternative. It has been a common experience that in a bid to undertake a research program, a researcher spends somewhere between 20-80% of his/her time in the procurement of goods and making the experimental setup functional. Generally, such infrastructure remains inadequately utilized because either funds dry up to continue using it and/or researcher moves on to some other project. Another researcher, after a gap of few years, working in similar research domain would seek similar infrastructure again for his/her work, consigning the earlier infrastructure to junk. Difficult procurement process, inadequate utilization, and post-project non-availability to other users are the reasons to look for better options to improve the existing process of conducting research from the points of view of the funding organization as well as researchers.

Rationale

We have witnessed situations where within one small geographical area, multiple public-funded research facilities of similar kind were created, not because the number of researchers was large but because the administrative units were different. The scenario is similar to that of farmers in a village rushing to buy tractors as a status symbol whether or not they actually had the need for so many of them. Time has proven that such investments have not been sustainable and several of these facilities got eventually shut down.

Then, there are situations where the shortage of funds is known beforehand and expected utilization of the proposed infrastructure is large. In such situations, common research facilities are created. Establishment of National Phytotron Facility (NPF) at IARI, New Delhi is a case in point. Researchers at different locations in the country wished to grow their plant material under controlled environmental conditions for a variety of reasons. Creation of such facilities, including greenhouses, growth chambers and supporting infrastructure, at different locations was not possible due to exorbitant costs and dearth of technical know-how to build, operate, and maintain such facilities at multiple locations. Therefore, NPF was conceptualized as one such facility with pooled investments from ICAR, DST, and FAO. The NPF has largely served its role as a common research facility over the past two decades because ICAR committed adequate financial and human resource to operate it. It requires a lot of hard work, perseverance, administrative support, and utilization plan to make such common facilities successful in public sector. Institutional resources in public domain often become constrained at some point in time and such facilities suffer, sometimes, irreversibly.

The most important causes of research cost and time overruns and compromise on the quality of output are the strenuous procurement of instruments/ equipment and hardship in setting up the experiments in time. These difficulties practically dissuade researchers from undertaking challenging investigations. Obviously, the best performance of an agricultural researcher at present is seldom realized even after heavy investments. Alternately, if such researchers had their research logistics streamlined, the study would be completed in the expected time without cost overrun and with better quality of output.

Therefore, a need is realized to off-load some of the infrastructure creation activities by academic and research institutions and, instead, make the best use of the available services industry. This would create a more conducive research ecosystem at reduced cost and researchers would give their best research output.

Policy Paradigm

In view of the foregoing, it should be very convenient if common services for a researcher were available through private service vendors on payment. The researcher would then be acquiring only the essential key instrument(s), if at all necessary, and focusing on the core research issue. Reducing the hardship of procurement would minimize the drudgery associated with the setting up of the research infrastructure and also reduce the quantum of required financial resources.

Consider, for example, a food scientist who undertakes to develop a functional food from a set of raw materials for a given consumer group. Normally, the scientist would seek an access to a whole lot of laboratory instruments to be able to measure the properties and functional behaviour of raw materials and the constituted food. The researcher may have access to some of the instruments that exist in his/her lab. Funds would be sought for the remaining instruments required for the research. If the funds were made available, the procurement process would be initiated and, hopefully, the desired instruments would arrive in time. The research on the development of the functional food would then be pursued and completed. However, if the researcher (albeit capable) happened to be the one who did not have access to even one of the required set of instruments, such as in a developing institution, the funding agency would turn down the proposal due to heavy fund requirement for instruments. The requirement of setting up an elaborate research set-up, thus, would prevent a scientist in a developing institution from pursuing a worthy research.

Here is an alternate proposal. The research indicated in the previous paragraph would require considerably lower public funding and time allocations, if the set of instruments required for the determination of the relevant properties of the raw materials and the output were available, say, through a private food science laboratory on payment basis. No allotment of funds for instruments/equipment, human resource to operate the experimental setup, and consumables would be required. Instead, the researcher would only seek operational funds to meet the cost of outsourcing the specific services. Resultantly, the researcher is spared of the hardship of procuring the instruments/ equipment and relevant consumables. The researcher's time is now more productively utilized for timely completion of the research. Similar examples could be cited from diverse agricultural disciplines permitting the researchers to utilize the available time more efficiently for core research activity. The question now is whether such common facilities could be conceived and created.

One of the proposals is to set up enterprises or promote startups by the SAUs and research institutions jointly with private sector. The private sector partners may be agricultural science and/or technology graduates who have the comprehension and appreciation for the envisaged laboratories and test centres. The existing infrastructure could be spared by the institutions/ universities and additional funds may be made available by the Government or venture capitalists to make the Common Research/Testing Facilities Center (CRTFC) up-to-date. The technical human resource, initially, may come from the institutions and the universities on loan basis or permanent absorption basis. The rates for different services offered by the common research facility centers may be decided jointly. The center would have the flexibility and freedom to offer its services to the partnering institution as well as to other clients from industry and farm sectors. Soil health cards and precision farming, for example, need periodic up-dating of soil data which could better be undertaken by such CRFCs, providing high quality services without break. Agro-food processing units in production catchments would utilize the services of these CRFCs to ensure the quality and safety of their products.

Alternatively, an entirely new CRTFC could be established as a startup in purely private sector with the latest instrumentation and highly skilled operators to offer services to researchers and entrepreneurs alike. The CRTFCs are expected to be sound business enterprises beneficial to both, the institution and the private partner.

With private sector on board, Governmental institutions would be free from the procurement of lab instruments and equipment to a large extent and the resulting research portfolio would be more responsive to the contemporary needs. As a result, large inventories of unserviceable instruments and equipment would be done away with. The academic and research institutions would be able to trim down the human resource, which would otherwise be required to operate and maintain the research infrastructure. The academic and research institutions would maintain specialized facilities and research infrastructure not available in Common Research Facilities Centers. As a check on the labs in private sector, a few referral labs/ test centers in public sector may be maintained.

Besides agricultural research, farming today is also turning to be technology driven. Quality of inputs, soil health, performance of farming machines, and diagnostic tools need to be ascertained and certified to minimize the chances of poor crop and animal productivities. The existing public funded infrastructure is not able to cope up with the amount to testing required to meet the expectation of the stakeholders. Increasing level of farm mechanization is leading to the manufacture of newer and better machines. However, a mismatch of adequate testing facilities in the country is holding up the manufacturers from making these machines available to farm sector (Kesavan, 2018). A soil health card has become an excellent tool to let the farmer know the remedial measures that she/he needs to undertake to maintain the overall health of the soil. However, the soil health card needs to be updated regularly for reliability. Public funded infrastructure cannot meet the need to update the soil health cards of 130 million farm holdings. There is need for inviting the private sector to establish testing facilities to provide the necessary services in a time bound manner. The mission to issue Aadhaar cards in the country could not have been accomplished without the support of private sector. Government, of course, need to formulate effective regulatory mechanism, implement it, and monitor for ensuring the success of the mission.

A concern arises regarding the credibility of data generated through such outsourcing of testing jobs to private sector. Ministry of Food Processing Industries, GOI, has been funding the establishment of food testing laboratories in private sector to provide the services to food industries and food business operators. Similarly, DST, APEDA, and DBT have been encouraging the creation of research and testing infrastructure in private sector to provide services to industries, traders, entrepreneurs, business start-ups, etc. Government of India has the mechanism to grant accreditation to such laboratories. Many of these laboratories have earned good reputation at home and have been recognized by importing countries for the test results. It is clear that given the right incentives and regulatory framework, the private sector can provide effective and sustainable investments and innovations in the service sector and further encourage private investment in R&D and thereby contribute to higher agricultural growth (Singh, 2011).

There are several examples, today, of excellent services provided by relevant service vendors, e.g., limousine service, farm-machinery hiring service, medical diagnostic services, internet café, and so on, where the services are accessed by users today without having to invest in the infrastructure. On the other hand, the service providers make full utilization of the infrastructure to justify the investment. There are performance yardsticks for the service providers, facilitating a user to choose the one who best meets the requirements. Competition, accreditation, regulation, and monitoring have been helping in improving services offered by private vendors.

Such a paradigm for engaging private sector to offer a majority of testing services for R&D sector (Annexure A) would not interfere with or be detrimental to the intellectual property rights of the researchers and the institutions through developing effective protocols. Instead, the investments infrastructure and human resource would be more productive and competitive. Such private sector labs would operate 24 x 7 and also cater to the needs of clients from industry and farm sectors for sustainability.

While expenditure on research infrastructure would go down considerably as a consequence of adopting the proposed policy, the available budgetary support for agricultural R&D would be able to support many more number of programmes than before. This overall research facilitation would encourage youth to join academics and research leading to increased scientific temper in the country. Annexure II summarizes the major benefits that would accrue from such a policy intervention. The new paradigm would lead to public funds being spared from creating expensive research and testing infrastructure. At the same time immense opportunities for employment would have been created in the country giving boost to the national economy. Practically, all developed countries rely on private sector for partnering with public funded institutions for meaningful R&D as well as testing services.

Conclusion

At present, huge public financial resources and scientific efforts are required to create research infrastructure, which is not utilized efficiently. A policy may be put in place to encourage, support, and incentivize private sector, either in PPP mode or otherwise, to create, maintain, and provide high quality testing services to agricultural R&D institutions in accordance with the highest global standards of laboratory practices. The policy would enable the agricultural R&D

sector to considerably reduce its time and money investment in the creation of avoidable infrastructure and, thus, minimize its inertia so as to better respond to national priorities. At the same time, these laboratories and test centres would enable the farmers, entrepreneurs, and traders to access the testing services conveniently. Thus, a new window for service-based startups would be opened leading to creation of additional employment opportunities.

References

Byerlee, D. and Alex, G.E., 1998. Strengthening national agricultural research systems - policy issues and good practice (English). Environmentally and Socially Sustainable Development Series. Rural Development. Washington, D.C. The World Bank.

FICCI, 2017. Harnessing private sector investment in R&D. Federation of Indian Chambers of Commerce & Industry, Federation House, Tansen Marg, New Delhi-110001. 52 pages.

Kesavan T.R., 2018. Personal communication. Chief Operating Officer, TAFE motors and tractors Ltd., Chennai.

Singh, A., 2011. The changing landscape of public expenditure and investments in Indian agriculture: Implications on growth trajectory. Ind. Jn. of Agri.Econ. 66(3): 301-313.

Annexure I

Indicative services which could be outsourced to private service vendors

- Characterization of raw, partially, and fully processed materials
- Soil, water, and leaf biochemical analyses
- Quality testing of farm inputs
- Plant and animal pathological samples' testing
- Assessment of physical, biochemical, and microbiological contamination through already known diagnostic protocols
- Biotechnological protocols for newly designed life forms
- Controlled environment facilities to study growth of agriculturally relevant life forms
- Identification of pests and diseases
- Pesticide residue determination
- Testing of machinery for farming and post-harvest operations

Annexure II

Possible benefits from the new policy paradigm

- Reduction in public expenditure on institutional research infrastructure
- Reduction in consumables and operating expenditure related to the reduced institutional research infrastructure
- Reduction in human resource and logistics for handling the reduced institutional research infrastructure
- Convenient and timely execution of research programs
- Harnessing the strength of private sector for organization of services and infrastructure
- Greater participation of agricultural researchers in research programs due to better research ecosystem
- Creation of self-employment and business opportunities in providing agricultural research and testing services
- Expansion of agricultural research portfolio with the allocated research funds.

Chapter **6**

Strategies for Market Orientation of Agri Startups for Smart Farming

P. Adhiguru[1], and G. Letha Devi[2]

[1] Principal Scientist, Agricultural Extension Division, ICAR, New Delhi
[2] Senior Scientist, ICAR-NIANP, Adugodi, Bengaluru
E-mail: padhiguru@gmail.com

Agricultural Market Ecosystem

Agricultural market environment at national level and international level keeps changing in various ways over the years. The market dynamics affect farmers, especially small farmers as the traditional subsistence farming no longer can provide support for increasing standard of living. Also, the rural youth look for other forms of employment and modern ways of doing agriculture. In this context, increased engagement in market-driven agriculture is becoming the reality for most of the farming community.

Market-driven agriculture has the scope to provide opportunities to farmers, provided, the farmers are up-to-date with market intelligence and they could meet out the desired quality standards in farm produce. However, small farmers, agriculture related traders and ago-based processors are experiencing difficulties to avail the emerging opportunities due to increased transaction costs and inefficient value chains for their products. Some of the constraints small farmers experience in taking advantage of market opportunities include meeting demands of markets in terms of quality and quantity standards and timeliness in operation and marketing of the farm produce at right time. More, the farmers are experiencing the necessity to shift from production for predominant home consumption to predominantly meeting the market demands. In such a scenario, agri-startups has tremendous scope to play to provide quality agriculture advisory services.

There are observations that some of the Agri-startups which offer various advisory services are unable to support the clients i.e., agricultural producers'/business actors in value chains for agricultural products and also support effectively in increasing their competitiveness. Such situation arises due the fact that some of the personnel engaged in Agri-startups do not receive appropriate training and also inadequacy in research towards market orientation of services. Therefore, majority of Agri-startups need to develop their understanding of various market opportunities, if they are to achieve more market success. Market-oriented agricultural advisory services can play a vital role in facilitating their clients to overcome know-how constraints and other farming needs.

Market orientation of Agri-startups include a highly diverse range of components ranging from technical know-how, understanding of markets, their requirements and business management to organizational development, and improving value chain efficiency. This market orientation ensures sustainability of Agri-startups. The broad components of Market orientation in agriculture may focus in areas of Production, Processing and Marketing. This may include the following:

The Agri-startups need to have thorough understanding in terms of technical know-how for improving quality, quantity and timing of production for ensuring better marketability which also includes Good Agricultural Practices (GAPs). The Good Agricultural Practices in this context, may include selection of appropriate commodities, varieties in the case of crops, breeds in the case of animal and fishery related enterprises which are suitable for the market, conducive agricultural management practices may be like soil fertility management, plant protection and water management.

For transforming from sustainable agriculture towards Agri-business, the startups necessitates know-how on farm management economics, business management tools and markets. Specific requirements for startups include farm enterprise analysis, marketing, market analysis, business planning and record keeping, advice on legal, regulatory and certification issues.

Efficient value chain management is crucial for successful agri startups and therefore, know-how about enabling stakeholders across value chain to attain standard quality requirements of the farm produce, for better marketability is important. Emphasis of the enterprise need to be on converting farm produce to marketable product. Value addition not necessarily mean only processing; it may be in the form of time saving/space saving/money saving/better form/improving the overall efficiency of the value chain. These may include Post-harvest management practices, improvised storage for enhancing shelf life, quality specification on products, quality standardization, certification, labelling, branding, use of model packaging technologies, ensuring food safety and phytosanitary standards by good agricultural practices, and awareness on consumer rights and regulations. Further, there is a need for wider technological applications in the area of sorting, grading, increasing shelf life of agriculture produce (fruits, vegetables, flowers).

For encouraging Agri-startups having ideal market orientation, there is a need for capacity building of stakeholders and value chain actors across the value chain is crucial. The orientation of startups need to be in the subject matter areas like management of financial aspects, enhancing leadership qualities and solution-centric negotiation skills, situation analysis and handling, strategic planning and participatory innovation development and management.

Appropriate system need to be evolved at macro level for market orientation in Agri-startups for enabling needed changes in improving value chain management in the way of better co-ordination of production processes, promotion of collective marketing, agreeable contract negotiations, legal formalities and procedures, establishing and promotion of branding, bridging Agri-statups with various supply chains to different marketing platforms including supermarket chains or fair trade, organic and other specialized markets, awareness creation and facilitations for accreditation to concerned agencies and their farm produce certification.

Another key aspects of market orientation include facilitating linkages among various business actors along value chains like convening multi-stakeholder meetings or groups to have up-to-date knowledge on market trends and drivers of growth, to foster better mutual understanding and trust, to identify bottlenecks along value chains and devise solutions, and to assist traders and processors to link up with reliable producers.

Agri-startups can focus on different type of services viz., input supply, finance, insurance, packing, aggregation of products, storage and transport logistics are needed in order for farmers, traders, processors and other value chain. Agri-startups engaged in value chain approaches need to join together with other stakeholders to facilitate provision of market driven set of integrated set of services and infrastructure to meet the needs of clients.

Types of Agri-startups who provide agriculture-based services

Producer and commodity startups: These startups may be in the form of village level commodity specific producer groups, regional level producer associations and co-operatives, processing and export associations. In general, such startups focus on providing services related to marketing and also, sometimes provide technical know-how services by engaging their own subject matter advisors or through linking with external advisors. Such services may help producers and other stakeholders in value chains to improve cost efficiency and increase the net profits.

Processing and trading startups: Processing and trading enterprises usually provide services to ensure they get the right quality and quantity of produce at the right time from their suppliers. These services often require more specialization and are usually associated with contract farming and other agri-businesses.

Independent private service startups: These startups in the form of individuals, small firms, training and resource institutions usually provide business development services rather than mere providing of know-how services. Also, private startups mostly focus on high potential areas where the profitability is assured.

Input supply startups: They focus on suppliers of seeds, agro-chemicals, equipment and to animal feed, veterinary medicines and equipment and also provide technical information related to queries raised by producers.

Market orientation gaining relevance

For providing efficient service, and making it attractive to clients with high quality, the Agri-startups need to thorough understanding on know-how and maintain high quality standards. The personnel engaged in Agri-startups need to have periodical capacity building and get updated with regard to changing market scenario. If the startups do not have dynamism in market orientation, quality of their service is likely to come down, and any kind of irrelevant or faulty market advice to the clients across the value chain clients may seriously affect the credentials of the startups. Therefore, startups need to have inbuilt training and mentoring in technical, economic subject matter and in the area of facilitation and methodological skills, exploring and testing new technologies. Agri-startups by providing more transparent advisory services with appropriate market orientation can make a contribution towards meeting need of the farmers and actors along value chain there by help for getting enhanced profitability to them. Also, such a market-oriented service is essential component for inclusive agricultural development.

Market-orientation need to have research base

Drawing lessons from research findings related to marketable issues will be an important for market-orientation of startups and providing services by them.The startups for improving the services need to venture into more of technical innovations including use of information and communication technologies along the value chains and also, they need to explore the possibilities of evolving social and institutional innovations which may help to bring synergy among actors in the value chain, moving products to market more efficient way, ensure competitiveness and profitability, and establish linkages among producers, processors and traders.

Recently, the National Agricultural Research Systems are started focusing on carrying out research suited to the market demands in the subject matter areas, such as the market system and market intelligence analysis, institutional arrangement of market institutions and their operational mechanisms. In this context, a synergy between researchers engaged in such topical researches and those engaged in various startups would be of much beneficial for promoting market orientation of startups. Also, the agricultural researchers also can learn from Agri-startups with regard to multiple dynamics of Agri-startups ecosystem. Agri-startups can fine tune their strategy based or research findings towards market-orientation of their approaches and services. Such kind of synergy in conducting research and sharing of research findings would also be useful for the Agri-startups to deliver services which can effectively address environmental issues, food security concerns, food safety standards, etc. In addition the market driven research can focus upon the risk minimization with regard to sustainability of startups which may be due to environment, changing consumer preferences and market volatility in prices. For this purpose, the researchers need to have good understanding on various issues concerning crop-specific value chain and also, the needs of stakeholders in value chain including startups.

Market-orientation to focus on all actors in value chain

Agri-startups need to focus their market orientation in such a way that their product and know-how advisory services need to go beyond only farmers, rather need to work with various actors in the value chain or any market setup which can provide efficient solutions for overcoming the constraints and enabling to achieve benefits with respect to the specific role played by each actor in the value chain.

Need for change in mindset

Startups usually focus on mere farmers, instead they need to focus on wide range of stakeholders for their sustenance and improved efficiency of services. Also, conventionally there is lack of trust between those stakeholders in Government with private sector players like traders, enterprises engaged in processing and logistics and market intermediaries. Policies need to recognize these multiple actors who are associated with markets and have potentials to facilitate resource poor producers.

Public sector organizations sometimes facilitate Agri-startups for market orientation through various interfaces in a project mode. However, such support should not be short-term, rather on sustenance basis because the market environment keeps on changing over the time. Also, such approach will ensure convergence among various actors across the value chain.

It is generally observed that resource poor producers have less risk orientation and on the similar line, even startups do face the risk one form or the other. A major role of public investment is, therefore, assumed to be one of motivating the poor to overcome this attitude. Uncertain and unreliable markets trends put them in a position of having less risk orientation. Especially in commercial mode has in built risks. Markets are always risky, especially in changing value chains and lead to risky situations. Appropriate support for startups in the form of knowledge and information for reducing the risks of increased exposure to market factors and facilitating them to make informed decisions in managing risks is needed.

There is a need to strengthen startups by providing them better understanding about nature of functioning of different markets; creating awareness about opportunities available in a market-oriented environment. Startups need to have abilities for accurate assessments of economic potentials and the possible risks of various products and enterprises for their marketability. They need to understand the methods to facilitate increased transparency regarding prices, regulations and standards. Also, need to be engaged in facilitation of multi-stakeholder platforms to openly discuss the interests of different actors. Startups need to gain capacity to support producer in enhancing their negotiating skills. One of the market-orientation startegies in risk minimization perspective for startups is to diversify the service for mix of products with adequate quality which are suited to markets with high demands.

Some market-oriented approaches for risk minimization

Insurance: One of the tool for mitigation of risks related to climate variations like droughts and floods, and the startups can offer advisory services to overcome such eventualities by producers. **Warehouse receipt systems:** Such systems reduce producers' exposure to price fluctuations and can help for proper household cash flow by ensuring access to storage of the produce. **Water management:** Water management is one of the key area for climate-related risks. This helps for efficient input management and ensure timeliness, quantity and quality of produce, especially for high value farm produce.

Enabling actors in value chain

In general, producers and other value chain actors hesitate to take the risk of more market-oriented strategies. But there is emergence of new market actors across the value chain with need for quality standards and specifications, product certification and procurement methods. Startups has scope to play a role by providing services with regard to know-how about such market pathways, legal and regulatory procedures. Another dimension by which startups can effectively play an important role is promoting trust and transparency in relation to alternate emerging models like contract farming. The startups apart from providing technical know-how can facilitate synergy between producers and commercial entities in the case of contract farming, which can help reducing risks for small-scale producers. Market orientation of startups is pre-requisite, if they want to venture to delivery of services to new institutional arrangements like contract farming.Basic understanding on legal aspects is very much integral part of contract farming. Therefore, startups can extend services for providing legal advices and protocols to ensure transparency about 'terms and conditions', which can help producers better negotiate and harmonize with the commercial entities with whom contract farming arrangement is made.

Market-orientation vs Gender concerns

There are observations that as commercialization of agriculture increases, access of farm women to income, sometimes decrease as such commercialization lead to greater role to men. Thus, when a shift is happening from subsistence farming to high value agriculture or cash crops, the role shifts to men. Such a situation also influences the income flow to women and decision-making power of women towards negatively. In view of this, when startups design their services towards market, orientation, they need to consider gender issues too. Market-orientation of Agri-startups need to consider potential for positive impact on women by various approaches like pro-actively promoting commercial opportunities for women, offering services specially tailored to women's needs and interests, employing an adequate number of women advisors and continuous gender-differentiated monitoring of outcomes and impact by developing gender-based indicators.

Market-orientation vs Food security

Household food security may get affected by greater market-orientation due to neglecting food production and insufficient allocation within the household of additional income to buying food, but it can also be due to failures in commercial activities leading to debts and loss of productive assets. Therefore, Government has greater responsibility to guide and monitor market-orientation of startups such a way that it minimizes the risks for negative effects of market-orientation on food security. Also, need to carefully monitor the impacts that commercialization has on resource poor producers and develop safeguard mechanisms that help actors from making losses in the commercialization process. It is observed that well-off startups are assumed to be able to develop their market relations with least government support, whereas those startups having less resources need greater support from government to avail viable market opportunities for them. Therefore, Government need to play crucial role in favour of weak startups in identifying market opportunities such a way that it generates greater potential for rewards in terms of generating sustained profits while rendering servicing to resource poor producers by startups.

Market-orientation to benefit pro-poor

Policy frameworks for pro-poor market-orientation need to consider the following: incorporation of market-orientation as an agenda in national policies like poverty reduction strategies, public investment frameworks, agricultural sector policy and plans and economic policies; there needs to be a comprehensive strategy for promoting market-orientation in the form of relevant investment in research, legal and regulatory structures, financial services, rural education and infrastructure; there need to be reforms in ministries of agriculture together with mechanisms for co-operation; market-orientation policies need to consider environmental considerations and the challenges and opportunities arising due to climate change; and procedures need to be in place to monitor the impact of market-orientation policies on different groups.

National level Capacity requirements to support market-orientation

Strong capacities are needed for monitoring and assessing macro-trends in agricultural market ecosystem. Various aspects that need to be considered are determining comparative advantages, identifying emerging threats to domestic markets, choosing investment priorities, and recognizing the changing landscape of opportunities and risks facing the poor. Apart from these, some specific factors need to be taken into account. These include the status of WTO, regional trade agreements, and changing trade barriers for different products; changes taking place in quality and food safety standards; the implications of growing demands for products, such as bio-fuels, fair trade and organic certified products, and livestock products; changing prices and emerging niches in global commodity markets; changing consumer food preferences at national and international levels; impact of urbanization and climate change.

Public-private synergy in investment for market-orientation

Public investment for market-orientation of startup needs to focus on where there are latent market opportunities to help resource poor producers, but where private investment is discouraged by risks and uncertain profits. Public investment in market-orientation need to be such that it fill up the gaps in service provision to actors across the value chain which also include developing the capacities of Agri-startups or private service providers to enhance the market access of resource poor actors. The public investments and private investments need to be complimentary to each other such that market-orientation of startups ultimately benefits resource poor farmers in agriculture.

There are considerable private sector investments in Agri-startups both input and output markets, especially in the area of agri-chemicals, food processing and modern retail. Public investment need to be in the area of creating conditions by which the private sector investment can be leveraged for more inclusive agricultural development coupled with having strengthend regulatory structure in place. Public-private partnership arrangements can be a practical way of combining public and private funding to enable Agri-startups and other actors in value chain to get enhanced profits. There is need for public investment for Agri-startups' capacity development and innovation.

Need to evolve market-orientation models

Sufficient efforts are needed to learn systematically from various projects in terms of lessons for policies and institutional development for effective market-orientation of startups. In recent past, there are many initiatives coming up to help value chain actors to respond to the changing market ecosystem. Many market linkage initiatives have no explicit 'theory of change' explaining how the approach could be expected to promote com petitive markets strategies for further scaling up to benefit more people. Hence, more efforts are needed to learning lessons both, from market linkages initiatives as well as from the growing wealth of experience regarding market-orientation of startups. Upto now, there has been very little appropriate evaluation of the experiences to determine whether outcomes from such market-orientation innovations have proven to be replicable or sustainable, and the extent of cost-efficiency.

Futuristic market-orientation

A conscious strategy for market-orientation is startup is essential for agricultural development keeping in mind the dynamism in national and international agricultural markets. Emerging factors like climate change, demands for bio-fuels, expanding dominance of modern retail are to be considered while prioritizing public investment for market-orientation of startups and also, attracting private investment in agricultural sector. Consistent efforts are necessary to support value chain actors to develop the knowledge and information for vibrant market-orientation of startups to match with changing consumer preferences and quality standards. Resilience to future changes requires both flexibility and clear long-term commitment to investing public resources to strengthen role of startups in value chains. Market-orientation need to be integral part of other approaches to decentralization, poverty reduction and risk mitigation. There is a need for new perspective to support the growth of private run startups with empirically proven impacts on resource poor producers.

Policy suggestions

Public policies for market-orientation need to be framed based on broad consultation of startups and other actors across the value chain. Startups, if strengthened, offer possibilities for multi-stakeholder dialogues on the best ways of developing agricultural markets to match with producer needs. Such public policies need to support startups to establish their own self-regulatory mechanisms in the form of product certification, ethical trade norms, etc., Conducive policies which are necessary for market- oriented startup to operate effectively such that to strengthen access to markets for the farmers, especially resource poor and promote sustainable market-driven agricultural development. Experience with such startups has shown the importance of greater attention to scaling-up, but has also shown that this need to be done within local and national processes of learning from experiences. Policies and priorities must be determined within a broader analysis of the actors and market dynamics in a given context.Governments need to enforce policies and legal structures; also create regulatory environment, support for certification or branding of quality products and fair competition across the startups. Government need to put that such that strengthen the position of the private sector in the market place, including a regulatory environment that market-orientation policies into action by ensuring that laws, norms and regulations are understood by the startups and other actors by facilitating optimal institutional arrangements.

Conclusion

A vibrant system is need of hour to address the issues concerning understanding of market, its trends and drivers, to strengthen better mutual trust building among stakeholders and to identify challenges in ensuring efficient value chain management and building appropriate strategies to facilitate effective coupling of various traders and processors with Agri-startups. There is a need to translate the lessons being learnt from various initiatives in market-orientation of startups, based on public and private investments so that such initiatives can be scaled-up with active support of policy-makers for the overall agricultural development.

❑❑❑

Chapter 7

Capacity Development and Skill Development Needs of Agri Startups

Bharat S. Sontakki and R. Praveen Kumar
Extension Systems Management Division, ICAR-NAARM, Hyderabad
E-mail: bharat235@gmail.com

Introduction

Agriculture, the world over, has been seeing rapid changes. While it continues to be a major subsistence livelihood option for majority in rural areas, it is also emerging as a fast growing business. It is becoming more scientific and efficient, with longer growing seasons and higher animal welfare and environmental requirements. It is not just regulatory processes, however, that will drive future farming. Success depends on meeting market requirements, driven ultimately by consumers and different trading conditions. Farming productivity is dependent on having a sustainable market for goods. Our farmers compete in national and international markets. These impact upon their farm practices, for example, how crops are grown and stored, the kinds of livestock kept and how they are reared. Farm profit is determined as much by cost base (such as seed, fertilizer, chemicals, fuel, breed, feed, machinery and labour requirements) as it is by prices paid. A good harvest in terms of national output can depress prices paid. Extreme climate change events or disease can have a devastating effect. The farming sector has to innovate and use technology where it can to do this, and this trend will be reflected in future planning applications.

Keywords: Agri Startups, Entrepreneurship, Capacity development needs, Capacity development framework, Competencies.

Entrepreneurship in vital for next generation agriculture

Traditionally, agriculture is seen as a low-tech occupation with limited dynamics dominated by numerous small family farms, which are mostly focused on production to meet household food and feed requirements. Over the last decade, this situation has changed dramatically due to economic liberalization, reduced protection of agricultural markets, and fast changing ecosystem. Farming practices increasingly have to adapt to the vagaries of the market, changing consumer habits, enhanced environmental regulations, new requirements for product quality, value chain management, food safety, sustainability, and so on. These changes are bringing new entrants, innovation, and portfolio entrepreneurship in farming. It is recognized by policy-makers, practitioners as well as scientists that farmers and growers increasingly require entrepreneurship, besides sound management and craftsmanship, to be sustainable in the future (McElwee, 2008; Pyysiäinen *et al.*, 2006). Recent studies show that agricultural entrepreneurship is not only wishful thinking or a new hype but also it has a profound impact on business growth and survival (Lans *et al.*, 2011; Verhees *et al.*, 2011).[2]

Agri Startup: *the new wave of entrepreneurship in agriculture*

Driven by technology and policy, startup culture is steadily growing in India. Start-up is an entity that develops a business model, based on either product innovation or service innovation and makes it scalable, replicable and self-reliant.

Startup is an early stage of entrepreneurial venture backed up by innovative concept or idea. Farm sector too is witnessing considerable startups. In addition to offering support to enhance the quality and yield of crops and other farm produce, the Agri-startups provide farmers with knowledge and extension services, geared towards improved planning and productivity, efficient adoption of scientific practices as well as stabilization of payments and finances.[3]

---

1 https://www.nfuonline.com/cross-sector/rural-affairs/planning-and-local-authorities/planning-news/how-agriculture-is-changing-technology/ accessed on 27 December, 2018

2 https://www.researchgate.net/publication/260985365 accessed on 27 December, 2018

3 https://inc42.com/features/watchlist-agritech-startups-2018/ accessed on 30 December, 2018

A startup is a company that is just beginning to develop and generally has the following features :

- Generally small
- Initially financed and operated by a handful of founders or one individual
- They offer products and services that is not currently being offered elsewhere in the market, or that the founders believe is being offered in an inferior manner
- In the beginning, expenses are more than the revenues
- Because of more expense, they require financing
- They can be funded from different sources[4]

Agri-startups, if promoted, nurtured and mainstreamed properly, can contribute immensely to agricultural development. They can:

a) Enable gainful engagement of YOUTH in agriculture,

b) Create jobs on farm and rural areas,

c) Attract investment in agriculture, and

d) Offer end-to-end solutions and services to farmers and consumers.

Startup ecosystem and current status of startups in India.

The startup ecosystem is the broad canvas representing different systems/actors that are critically important for creation, nurturing, promoting and sustaining startups. This ecosystem as illustrated in figure 1, consists of the actors like: i) idea creators, ii) resource providers, iii) educational opportunities, iv) funders, v) media and communication, vi) advocates or promoters and vii) sources of risk capital.

4 https://www.slideshare.net/Sahil Swangla/startup-india-agri-startups accessed on 30 September, 2018

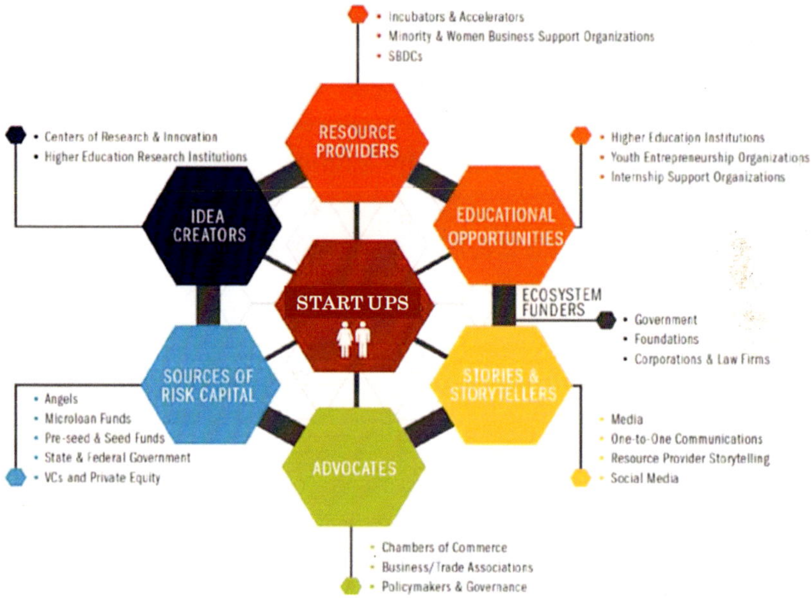

Fig. 1 : The startup ecosystem

(*Source:* modified from https://www.slideshare.net / Sahil Swangla / startup-india-agri-startups accessed on 30 September, 2018)

The Indian Government is supporting Agri-startups through its flagship programs like 'Startup India' and 'Make in India' programmes. India has enviable ecosystem for promoting startups for faster and inclusive growth in agriculture and sustainable livelihood improvement of farming communities. With demographic dividend due to substantial youth population, raising literacy and educational levels, expanding private investment, enabling government policies, it has increased FDI inflow by 33% and also, created easy ecosystems for startups. India has about 300 business incubators/accelerators, several programmes to provide entrepreneurship skills, offering seed funds for starting a business, grants and subsides, easy loans and simple registrations.

Nationally, a total of 366 agri-based startups have come up from 2013 to 2017 with more than 50% of the startups in the last 5 years started in 2015 and 2016. It is also to be noted that more than 90% of all funding is focused on seed stage and early stage startups.[5]

---

5 http://ficci.in/spdocument/23049/Agri-start-ups-Knowledge-report-ficci.pdf accessed on 29 September, 2018

Capacity development

Capacity development is the process through which individuals, organizations and societies obtain, strengthen and maintain the capabilities to set and achieve their own development objectives over time. Capacity is about growth of the individual in knowledge, skills and experience. Growth of the group that surrounds this individual, as these skills and knowledge are passed on. And from this individual and group, growth of a society and nation. Capacity development is about supporting growth-within individuals, groups and across societies as a whole.

Consensus is emerging, from the development community worldwide, that capacity cannot be separated from sustainable human development. There is also growing understanding that capacity must be viewed from three distinct but related perspectives viz., i) *Individual*: the skills and knowledge vested in individuals, communities and groups, ii) *Institutional:* the internal policies, systems and strategies that enable an organization to operate and to achieve its goals and iii) *Ecosystem:* the wider society within which individuals and institutions function.

Stages in the startup process and implications for capacity development

Process of startup consists of six distinct stages, namely: i) ideating, ii) concepting, iii) committing, iv) validating, v) scaling and vi) establishing and sustaining (Figure 2). These stages require different competencies and capacity development needs.

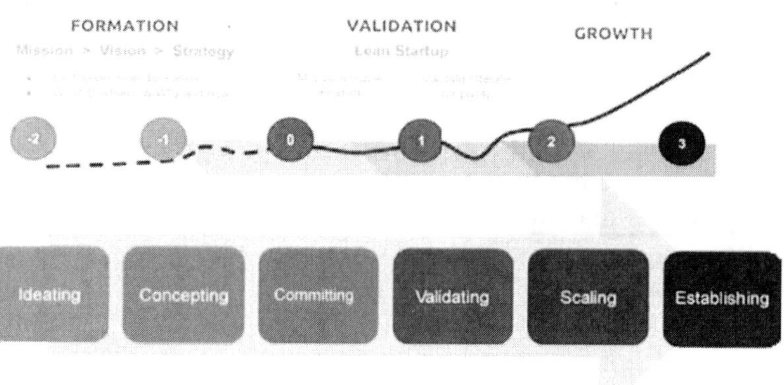

Fig. 2 : Phases in the startup process

(*Source:* modified from https://www.slideshare.net/Sahil Swangla/startup-india-agri-startups accessed on 30 September, 2018)

Ideating: This stage is an individual forms a potential scalable product/service idea for big enough target market and an initial business model on how it would make money. This state is characterised by one person or only vague team; no confirmed commitment and/or no right skills balance in the team structure yet.

Concepting: In this stage, the prospective startup entity develops a clear and meaningful target with clear direction for a minimum of three years with milestones to get there along a timeline of say, up to 3, 6, 12, 24 and 36 months. At this stage the team starts to evolve with two or three core founding individuals with balanced ownership. The team can also already have some extended team with lighter commitment (stock options and/or cash compensation).

Committing: In this stage, the startup team evolves considerably with clear commitments and skills balanced among them. The team is able to develop the product/service (Minimum Viable Product) without dependency of uncommitted external resources or may already have initial product/ service developed. The team at this stage has a signed shareholder agreement between founders, with milestones, committed time and money usage.

Validating: The startup at this stage shows some user growth and/or revenue (initial traction). Or, may continue to attract additional resources (money or sweat equity) for equity or future revenues. It starts looking for clear market validation (Product Market Fit), to be able to move in to scaling.

Scaling: Showing clear, growing and measurable user/market traction in big or rapidly growing target market. Can and want to scale fast and/or is able to attract significant funding.

Establishing: Achieved great growth, which can be expected to continue strong. No longer need to "try" to get resources and can get those easily. Continue to grow and often wants to culturally continue behaving like a "startup" for as long as possible. Founders make exit or continue business as usual.

Table 1: Stages in the startup process vis-à-vis capacity development[6]

Phases	Features	Capacity development needs	Approaches
Ideating	Entrepreneurial ambition and/or potential scalable product or service idea for a big enough target market. Initial idea on how it would create value. One person or a vague team; no confirmed commitment or no right balance of skills in the team structure yet.	Brainstorming Creativity Originality Decision tree[1]	Awareness/ sensitization programmes Online learning Boot camps Experience sharing Workshop with successful startups.
Concepting	Defining mission and vision with initial strategy and key milestones for next few years on how to get there. Two or three entrepreneurial core co-founders with complementary skills and ownership plan. May be additional team members for specific roles also with ownership	Team dynamics Interpersonal dynamics Visioning and strategizing	Formal training (Tinkering lab, Incubation) Accelerator Exposure visits Internship Testimonial/case history analysis[2]
Committing	Committed, skills balanced co-founding team with shared vision, values and attitude. Able to develop the initial product or service version, with committed resources, or already have initial product or service in place.	Communication and negotiation Market analysis Enterprise management Marketing management	Formal training On-the-job learning Network workshops
Validating	Testing assumptions for validated solution to demonstrate initial user growth and/or revenue. Initial Key Performance Indicators (KPI's) identified. Can start to attract additional resources (money or work	Domain clarity Quality management Finance and investment management	Formal training On-the-job learning Flawed Testing Methods Situational Testing Methods

	equity) via investments or loans for equity		Ideal Testing Methods[1]
Scaling	Focus on KPI based measurable growth in users, customers and revenues and/or market traction & market share in a big or fast growing target market. Can and want to grow fast. Consider or have attracted significant funding or would be able to do so if wanted. Hiring, improving quality and implementing processes	Enterprise management Risk assessment and management Customer orientation Branding and brand promotion	Formal training On-the-job learning Network workshop Accelerator
Establishing	Achieved great growth that can be expected to continue. Easily attract financial and people resources. Depending on vision, mission and commitments, will continue to grow and often tries to culturally continue like a startup Founders and /or investors make exit(s) or continue with the company	Business expansion Corporate social responsibility Media management	Formal training On-the-job learning Network workshop Policy engagement Social engagement

6 https://www.slideshare.net/SahilSwangla/startup-india-agri-startups accessed on 30 September, 2018

7 https://en.wikipedia.org/wiki/Ideation_(creative_process) accessed on 2 January, 2019

8 https://www.corwin.com/sites/default/files/upm-binaries/7647_Altstiel_Chapter_4.pdfaccessed on 5 January, 2019

9 https://medium.com/swlh/how-to-start-a-startup-e4f002ff3ee1 accessed on 4 January, 2019

Capacity development needs of Agri-startups

There is paucity of literature on 'capacity development and skill development needs of Agri startups'. Hence, we relied on literature pertaining to capacity development needs and skill development needs of agripreneures. After literature search, we chose to analyze this information in terms of competency needs of Agri startups. Competency is a more contemporary term and even the National Training Policy, (2012)[10] is premised on competency-based approach. Accordingly, the specific competency needs for Agri-startups are:

i) Domain competencies (in agri/smart farming/IT): Studies by Venkattakumar *et al.* (2014)and Rathore *et al.* (2018) reported that strong domain competencies like knowledge, understanding and skills in the specific farming area is a critical factor for entrepreneurial success in agriculture. Most of the emerging startups are based on leveraging technologies like geo-spatial, sensor, artificial intelligence, product engineering and quality management to make farming more efficient, effective, remunerative and consumer centric. Since these tech startups are in one or the other segment of agriculture like crops, animals, food, feed, supply and value chain management, a comprehensive understanding of the techno-legal aspects in these areas is a basic precursor. A startup without critical understanding, knowledge and skills in agriculture stands less chance of success.

ii) Business management competencies: These include competencies to start, protect and sustain the business. Competencies to manage customer, finances, business records, inventory management, and sales are essential and critical. Startups should also have competencies to promote products and services of the business, competencies to manage human resources, competencies to manage the business. Organizations are facing different business growth and development challenges; hence, the concept of producing greater entrepreneurial competencies and their activities has become a prominent goal, for many stakeholders, government and policy-makers. Entrepreneurial competencies seen as important to business development has been studied by many researchers (Davidsson, *et al.*, 2006).[11]

---

10 http://persmin.gov.in/otraining/NationalPolicy.pdf accessed on 6 January, 2019

11 http://shodhganga.inflibnet.ac.in/bitstream/10603/136086/14/14%20chapter-4.pdfaccessed on 7 January, 2019

iii) Communication competencies: This category of competencies relates to person-to-person or individual-to-group based interactions, e.g., building a context of co-operation and trust, using contacts & connections, persuasive ability, communication and interpersonal skill (Man *et al.*, 2002) and to become successful entrepreneur needs to possess competencies in relationship building, communication, persuasive and interpersonal abilities (McClelland, 1987; Lau *et al.*, 2000). Bird (1995) also described this relationship building activities as entrepreneurial bonding.[12]

iv) Networking competencies: The emerging trends in knowledge-based economies and global markets have increased the need for network building for the survival of new businesses (Huggins, 2000). Network building is an effective instrument by which entrepreneurs improve access to business ideas, capital, knowledge and technology to generate innovative business start-ups (Zarefard & Cho, 2017; Aldrich *et al.*, 1986; Hatala & Fleming, 2007). The ability to build a large network of social contacts can empower entrepreneurs to obtain the necessary resources for their start-ups to prosper and grow (Liao *et al.*, 2005).[13]

v) Marketing competencies: These include competencies to: a) understand the nature of business, b) comply with Government regulations, c) deal with the business, d) finance the business, e) locate the business, f) plan the marketing strategy, g) choose the type of ownership, h) obtain technical assistance, i) develop a business plan, j) determine the potential as an entrepreneur.[14]

vi) Customer-orientation competencies: A customer-orientation has long been advocated as a business philosophy that leads to superior performance and firm profitability. The concept of a customer-orientation has been used synonymously with the term 'market orientation' and it has been operationalized as a dimension of a market-orientation construct (Narver & Slater,

---

12 http://shodhganga.inflibnet.ac.in/bitstream/10603/136086/14/14%20chapter-4.pdf accessed on 7 January, 2019

13 https://www.abacademies.org/articles/Entrepreneurs-managerial-competencies-and-innovative-start- up-intentions-in-university-students-focus-on-mediating-factors-%201939-4675-22-2-141.pdf accessed on accessed on 7 January, 2019

1990). Recognizing the distinct nature of a customer-orientation from that of the broader market-orientation construct, a growing number of studies have examined customer-orientation as an isolated construct. Moreover, a customer orientation, with its traditional focus on 'listening' to customers, has its share of critics who cry out the detrimental effects of a customer-orientation on organizational processes and performance (Christensen *et al.*, 2005). For these reasons, the customer-orientation construct has been undergoing conceptual debate and empirical validation (Herhausen, 2011).[15]

Need for developing 'Learning and Capacity Development Framework' for Agri Startups

The basic precursor for capacity development in any sphere is 'capacity development needs assessment'. This could be done by a multi-pronged approach using task/role analysis, competency mapping, surveys, Focus Group Discussions, and backed up a peer review workshop to evolve Learning and Capacity Development Framework. This framework should suggest a matrix to delineate Learning and Capacity Development approaches and methods, the relevant institutional arrangements, facilitating arrangements, cost involved and outputs and outcomes as per the identified capacity development needs of startups at various stages of the process. The framework should also have provisions for building in formal, informal and non-formal learning at various stages of startup process. The framework so developed should be reviewed once in every five years to continuously innovate on capacity development strategies.

Conclusion

Capacity development has to be approached as a continuous intervention during the startup process with various approaches like formal training, online training, informal and non-formal learning,

---

14 http://shodhganga.inflibnet.ac.in/bitstream/10603/136086/14/14%20chapter-4.pdf accessed on 7 January, 2019

15 https://ac.els-cdn.com/S1877042814039147/1-s2.0-S1877042814039147-main.pdf?_tid = ff45a4fb-5954-44a6- a269-5fc6df76e91c&acdnat = 1546938977_fb97a7b2dcb538cbbf8995ae22a6dda2 accessed on 6 January, 2019

apprenticeship and internships, exposure visits, networking workshops and policy engagement. Further, capacity building has to be appropriately topped up with enabling measures like facilitation, handholding, mentoring, coaching, networking, policy lobbying and lifelong learning. The following recommendations are suggested for capacity development for Agri-startups to promote 'SMART Farming':

- Approach capacity development as an integral component of entrepreneurship process (not one-time intervention, but a continuous learning intervention)
- Develop 'capacity development framework' for Agri Startups
- Create centers of excellence and community of practice
- Catch them young - aggressive campaigning starting from primary level
- Facilitate transition of farmer innovations in to enterprises - inculcate startup culture.

References

Aldrich, H., Zimmer, C. & Jones, T., 1986. Small business still speaks with the same voice: A replication of the voice of small business and the politics of survival. The Sociological Review, 34(2): 335-356.

Cho, S. & Gumeta, H., 2015. Factors affecting university students' start-up intention: Comparative study on Korean and Chinese universities. Journal of Korean Entrepreneurship, 10(3): 150-152.

Christensen, C.M., Cook, S. & Hall, T., 2005. Marketing malpractice: The cause and the cure. Harvard Business Review, 12: 74-83.

Chris Phelan, 2014. Understanding the farmer: An analysis of the entrepreneurial competencies required for diversification to farm tourism. Ph.D. Thesis submitted to School of Sports, Tourism & the Outdoors, University of East Lancashire, p. 313.

C. McClelland, David, 1987. Characteristics of Successful Entrepreneurs. The Journal of Creative Behaviour. 21. 219-233. 10.1002/j.2162-6057.1987.tb00479.x.

David Kahan, 2012. Entrepreneurship in farming. FAO, Rome. p.136.

Davidsson, Per, Delmar, F, Wiklund, Johan, 2006. Entrepreneurship and the Growth of Firms 10.4337/9781781009949.

Hatala, J.P. & Fleming, P.R., 2007. Making transfer climate visible: Utilizing social network analysis to facilitate the transfer of training. Human Resource Development Review, 6(1): 33-63.

Herhausen, D., 2011. Understanding Proactive Customer Orientation: Construct Development and Managerial Implications, Doctoral Dissertation, University of St. Gallen, Switzerland.

Huggins, R., 2000. The success and failure of policy-implanted inter-firm network initiatives: Motivations, processes and structure. Entrepreneurship & Regional Development, 12(2): 111-135.

Liao, J., Welsch, H. & Tan, W.L., 2005. Venture gestation paths of nascent entrepreneurs: Exploring the temporal patterns. The Journal of High Technology Management Research, 16(1): 1-22.

Man, T. and Lau, T., 2000. Entrepreneurial competencies of SME owner/ managers in Hong Kong service sector: A Qualitative Analysis, Journal of Enterprising Culture, 8(3): 235-254.

McElwee, Gerard, 2008. A taxonomy of entrepreneurial farmers. International Journal of Entrepreneurship and Small Business - Int J Enterpren Small Bus. 6 October, 1504/IJESB.2008.019139.

Narver, J.C. & Slater, S.F., 1990. The effect of a market-orientation on business profitability. Journal of Marketing, 54(4): 20–35.

National Training Policy, 2012 circulated vide Ministry of Personnel, Public Grievances & Pensions, Department of Personnel & Training O.M. No. 12021/8/2011-Trg.I dated 19 January, 2012 to all Central Ministries/Departments.

Pyysiäinen, Jarkko & Anderson, Alistair & McElwee, Gerard & Vesala, Kari, 2006. Developing the entrepreneurial skills of farmers: Some myths explored. International Journal of Entrepreneurial Behavior & Research. 12 October, 1108/13552550610644463.

S. Rathore, VKJ Rao & B.S. Sontakki, 2018. Documentation of Success Stories of Agri-graduates turned Entrepreneurs, 2018, Edn /2/11/ 2015-HRD Dated 28 July, 2017, ICAR-National Academy of Agricultural Research Management, Hyderabad.

Steven Carr & Anne Roulin, 2016. An exploration of Agripreneurship: Scope, Actors and Prospects. p. 71.

T. Lans a,", M.A. van Galenb, J.A.A.M. Verstegenb, H.J.A. Biemans a, M. Mulder a a Education and Competence Studies, Wageningen University, P.O. Box 8130, 6700 EW Wageningen, the Netherlands

Agricultural Economics Research Institute (LEI), Wageningen University and Research Centre, P.O. Box 29703, 2502 LS the Hague, the Netherlands.

Thomas W.Y Man, Theresa Lau, K.F Chan, The competitiveness of small and medium enterprises: A conceptualization with focus on entrepreneurial competencies, Journal of Business Venturing, Volume 17, Issue 2, 2002, Pages 123-142, ISSN 0883-9026.

R Venkattakumar and B.S. Sontakki, 2014. Impact of Agriclinics and Agribusiness Centres in Extending Broad-based Services to Farming Community. Project Report. ICAR-National Academy of Agricultural Research Management (ICAR-NAARM), Rajendranagar, Hyderabad. P (135).

Verhees, F.J.H.M. & Kuipers, Abele & Klopèiè, Marija, 2011. Entrepreneurial proclivity and farm performance: The cases of Dutch and Slovenian farmers. The International Journal of Entrepreneurship and Innovation. 12. 169-177. 10.5367/ijei.2011.0039.

Zarefard, M. & Cho, S.E., 2017. Relationship between entrepreneurs' managerial competencies and innovative startup intentions in university students: An Iranian case. International Journal of Entrepreneurship, 21(3).

❑❑❑

Chapter 8

Techno Economic Viability of Agri Businesses

Divya Rajput

Incubator Advisor to Assam Agricultural University, Jorhat

Email: rajput.divya @gmail.com

ABSTRACT

Agriculture is likely to emerge as the largest asset class and the next startup unicorn is predicted to be an Agri Startup (Source: Forbes, 2018). This is a global prediction and may not apply to Agri Start-ups in India, who are largely mission driven and not profit-driven. But why is profitability important and what does it imply for an Agri Startup?

Profitable businesses are relatively more sustainable, they have the bandwidth to serve marginalised stakeholders, highly empowered to make their own decisions, and attain greater operational autonomy. Profitable businesses can adopt technology faster than competition and innovate to achieve greater customer satisfaction. Profitable businesses which build an ethical value chain make great brands and create deeper value proposition for their customers. There are several stories of iconic brands, which have developed community initiatives and in turn their communities helped grow their profit margins. An interesting case is that of Tata Power DDN in Delhi, which turned losses to fabulous profits within a single decade, much to the credit of their community initiatives in slums of North Delhi.

One important aspect which a corporate understands while a start-up misses to ride upon is the techno-economical viability of business model. A corporate invests its resources only if the project or new business vertical makes perfect commercial sense. In this sense, a corporate is relatively risk-averse and sluggish at times to venture into new foray. The startups are built to launch new expeditions and mostly tend to solve a bigger market need or social problem. It is observed that a techno-economically viable project is more scalable, profitable and creates greater impact.

Keywords: Sustainable, Return on Investment, Viable business, Value Chain Integration.

The Author's Stint with Techno-Economic Viability

In 2016, as Head of Center for Business Innovation at India's premier Think Tank, Indian Institute of Corporate Affairs (under the Ministry of Corporate Affairs, GoI), we co-created an Agri-Tech Commercialization Platform with the intent to support Agri-Tech companies and start-ups jointly with the Technology Development Board (TDB is a statutory board under Department of Science & Technology, GoI) and Indian Council of Agricultural Research (ICAR under Ministry of Agriculture and Farmers' Welfare, GoI). TDB invests in technology ventures through debt/equity infusion if they meet compliance and are "techno-commercially viable". We scouted for innovative technology (indigenous or imported) driven ventures in agriculture which had the ability to double farmers' income. Initially, we laid focus on the pre-harvest domain for effective farm management and then reviewed allied agriculture, agri-residue management as well.

We investigated 60 + projects in 2016-17 and found 13 of them to be techno-commercially viable at the initial stage. I continued handling Agri-Tech portfolio for another 2 years with TDB and reviewed atleast 100 + projects only to realise that while most of them adopted technology, only few of them were able to relate technology with commercial viability. The good news is that most projects were socially or environmentally relevant.

What is Techno-commercial Viability?

One can find many definitions including what all parameters to calculate while assessing the Techno-Economic viability of projects but here is the simple language. One will need to assess how technology parameters impact financial returns while benefitting society and environment.

How adoption of a new process, machine, and/or service deliver better unit dynamics of the same agri-product? **Will the project meet all its financial liabilities (capital and operating expenses) and**

make enough to return to the investor and yet secure enough profits for business growth? What are the risks involved due to multi-stakeholder relations or dynamic nature of natural factors or changing market needs? What is the strength of the management team, the technical experts and what are their equations with their business? Who is accountable for technology or business model failure and who is the one to fix it or pivot, if required? And then let your financial statement speak for itself. Try and do this for every product you intend to work on and keep answering a basic question – "who will pay for using it?" Develop a revenue stream for every new product, whether operating in B-2-B or B-2-C space.

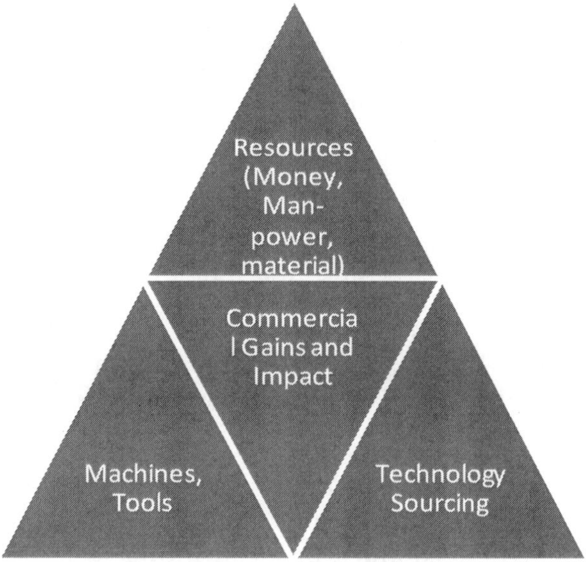

Fig. 1: Solve the Pyramid, investigate cash flows and propose profits

How does technology scale up ventures?

While we all like to hold the fanciest gadgets available, many of us may have realised that technology works around improving "user experience". Our user can be the farmer or the end-product user, anyone who will pay to use it.

The features and specifications of a technology, adoption of which leads to greater access of that tool, brings value addition and benefits to its users. Deeper market penetration leads to venture scale up. Availability of any two parameters may lead to a temporary gain.

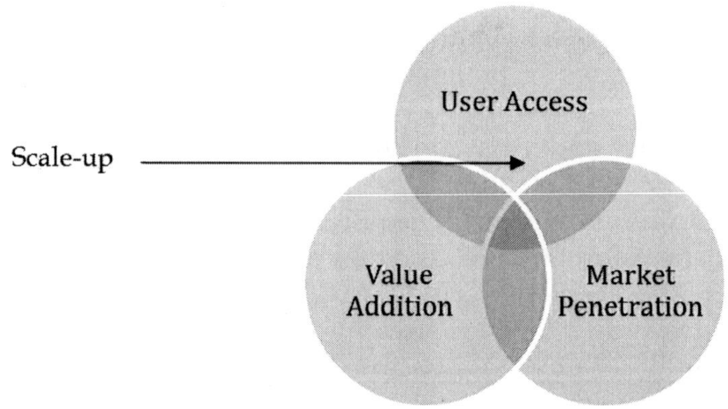

Fig. 2: What tools of your business model help you scale-up? Work it out.

Cost and Benefit Assessment

A. Costs

Costs include many visible and many hidden components; early stage start-ups with limited sectoral experience may not have full knowledge of all of those. Utilise the following parameters to work out cost assumptions:

1. Investment related costs; independent of operating costs. Mostly financed through own sources or debt funding through traditional mechanisms:
   i) Initial investment including planning (financing for construction, commissioning, operation permits, test operation phase or EPC contract) and consulting cost. This can range from 5-20% of the investment depending on the technology. *Most of the times, no one else pays for technology procurement.*

   ii) Administration and insurance costs; it ranges from 2-5% of the investment.

   iii) Infrastructure costs (building, land, license fees *et al.*)
2. Operational Costs: Mostly funded by private investors who charge an equity in return, sometimes through banks, at times as advance from a state agriculture department, which is keen to utilise its corpus to support working capital requirements of early stage startups. Incubated startups may like to utilise seed fund for working capital.

Includes manpower, labour, fuel, electricity, maintenance, consumables, business development or any other recurring costs. Include the important factor of working capital to procure raw material and supply chain expenses. This is where incorrect assumptions occur, so one must factor-in contingencies.

B. Benefits

1. Revenues from sale of products: consider availability of raw material, optimum usage of plant and machinery, ability to procure raw material and other consumables and most importantly, establish market linkages to sell those products.

   There must be atleast one core product/service/tool which brings most of the revenue while you are still working on your start-up!
2. Subsidies or grants: several state/central government schemes impart subsidies in agriculture sector. Find out about those and apply to get benefits.

C. Risk assessment

Startups must calculate risks involved in their projects by making relevant findings from similar on-going projects. While the commonly "known" risks can be categorised under financial risks, environmental risks, technical risks and social risks; the following para will lay focus on value-chain risks:

India has fragmented farm holdings with unorganised production capability, mostly rain-fed and climate change impact is affecting agriculture sector more than any other sector. In states which had organised production, water table and soil is polluted due to chemical over-use. Rural manpower is in migration mode as input costs of farming are higher than sale price and there is no buy-back except where market linkages have been established. Startups must carefully observe regions where they can build ethical value chains and strengthen local community through their initiatives.

Risks are mostly inter-related to each other. A start-up may like to consider the unexpected risk of technology failure or production breakdown which may lead to financial losses. It is observed that careful risk analysis may lead to alteration of business model at the right time. For example, you may like to outsource one of the process to ensure regularity than handle everything yourself. Such analysis can guide investments and contractual decisions. Increase in the raw material cost can alter the funding requirements. One may like to test the financial model to price variations and conduct a sensitivity analysis.

Inflation - It hits in true sense. In our part of the world, it is important to calculate the incremental costs due to inflation while integrating the value chain.

Common Terminologies

Startups will be required to make few assessments, such as below basis the requirements of the funder/ investor. One can learn about these terms in any relevant book or online literature or seek inputs from your CA or CS or Finance planner. In simple terms, *most of these assessments will help your investor to calculate the return on investment:*

- Static cost benefit assessment
- Annuity method
- Net cash flow table
- Net present value (discounted cash flow)
- Internal rate of return

Some of these are preferred over another due to ease of sensitivity analysis by varying few parameters.

"If you fear excel sheets and financial figures, hire a professional. If you are yet to reach at a stage of hiring a professional, recruit an intern – anyone who has dealt with accountancy principles or venture funding can work for you."

Food for Thought

Most of us are true technicians at heart. We love to create high precision, technical tools which not many others can understand. We add fancy names to it without realising that the other person doesn't understand its relevance. Ease is the name of the game; therefore, anything we create, must ease user acceptance, link to market needs and drive profit margins. Technology coupled with economic viability has the potential to create great value for its stakeholders. Simply remember that a profitable business allows you greater autonomy and that is the core faith you believed in when you became an entrepreneur!

❑❑❑

Chapter **9**

Agriculture 4.0: The Future of Farming and Agri-Tech

Sivakumar[1] and S. Suresh[2]

[1]CEO, MABIF and [2]Director and Nodal Officer, MABIF, TNAU, Madurai
E-mail: agribusiness.incubation@gmail.com

ABSTRACT

Technology is changing the world, and farming is catching up. The introduction of everything from automated farm equipment to a wide array of Internet of Things (IoT) sensors that measure soil moisture and drones that keep track of crops have changed the business of agriculture. Experts call this movement "Agriculture 4.0", a term used by the World Government Summit. The introduction of Agriculture 4.0 has produced a new term to describe those companies using this new and much more technified model: the Agritech sector. Basically, according to the European Parliament definition, it is "a farming management model, based upon observing, measuring and responding to inter and intra-field variability in crops". The goals are mainly increasing the productivity of the crops while ensuring a higher environmental sustainability.

Keywords: Agriculture 4.0. Hydroponics, Agtech, Bioplastics, Urban farming, Hydroponics, GHGs, Climate Change, Population Increase.

Introduction

The traditional approach of the food industry is undergoing a fundamental transformation. The first technology revolution in agriculture made impressive strides: Between 1961 and 2004, cereal yields in East Asia rose by 2.8 per cent a year, or over 300 per cent over the period, enabled by modern farming practices, including irrigation, use of fertilizers and pesticides, and the development of new and more productive crop varieties (World Bank, 2008).

But efficiency gains are dropping: The rate of yield increases has slowed. And the challenges are greater. The world has to produce 70 per cent more food by 2050, using less energy, fertilizer, and pesticide while lowering levels of GHGs and coping with climate change. Old technologies must be maximized, and new ones generated.

Agriculture 4.0, the coming agricultural revolution, must be a green one, with science and technology at its heart. Agriculture 4.0 will need to look at both, the demand side and the value chain/supply side of the food-scarcity equation, using technology not simply for the sake of innovation but to improve and address the real needs of consumers and reengineer the value chain.

Observations

A number of global trends are influencing overall sustainability of food and agricultural systems. The four core advancements placing pressure on agriculture to meet the demands of the future: demographics, paucity of natural resources, climate change, and food waste.

Increasing Population

According to the outcomes of the World Population Prospects 2017, the world's population amount to nearly 7.6 billion as of mid-2017, indicating that the world has added almost one billion inhabitants over the last twelve years. Sixty per cent of the world's people live in Asia (4.5 billion), 17 per cent in Africa (1.3 billion), 10 per cent in Europe (742 million), 9 per cent in Latin America and the Caribbean (646 million), and the remaining 6 per cent in Northern America (361 million) and Oceania (41 million). China (1.4 billion) and India (1.3 billion) remain the two most populous countries of the world, comprising 19 and 18 per cent of the global total, respectively.[1]

In the meantime, the global diet is varying too, as a result of ever-changing demographics: There's an upward demand for high-value animal protein, a trend that is being driven by urbanization and increasing incomes.

---

1 UN Population Division, 2017. World Population Prospects The 2017 Revision. [ONLINE] Available at: https://population.un.org/wpp/Publications/Files/WPP2017_KeyFindings.pdf. [Accessed 9 January, 2019].

Increasing Urbanization[2]

55% of the world's population lives in urban areas, a proportion that is expected to increase to 68% by 2050. Projections show that urbanization, the steady shift in residence of the human population from rural to urban areas, combined with the overall growth of the world's population could augment another 2.5 billion people to urban areas by 2050, with close to 90% of this increase taking place in Asia and Africa.

The 2018 Revision of World Urbanization Prospects produced by the Population Division of the UN Department of Economic and Social Affairs (UN DESA) notes that future increases in the size of the world's urban population are expected to be highly concentrated in just a few countries. Together, India, China and Nigeria will account for 35% of the projected growth of the world's urban population between 2018 and 2050. By 2050, it is projected that India will have added 416 million urban dwellers, China 255 million and Nigeria 189 million.

The urban population of the world has grown rapidly from 751 million in 1950 to 4.2 billion in 2018. Asia, despite its relatively lower level of urbanization, is home to 54% of the world's urban population, followed by Europe and Africa with 13% each. Today, the most urbanized regions include Northern America (with 82% of its population living in urban areas in 2018), Latin America and the Caribbean (81%), Europe (74%) and Oceania (68%). The level of urbanization in Asia is now approximating 50%. In contrast, Africa remains mostly rural, with 43% of its population living in urban areas.

Scarcity of natural resources[3]

Projections to 2050, suggest the emergence of growing scarcities of natural resources for agriculture (Alexandratos and Bruinsma,

---

2 UN Department of Economic and Social Affairs, 2018. 68% of the world population projected to live in urban areas by 2050, says UN. [ONLINE] Available at: https://www.un.org/development/desa/en/news/population/2018-revision-of-world-urbanization-prospects.html. [Accessed 9 January, 2019].

3 The future of food and agriculture | FAO | Food and Agriculture Organization of the United Nations. 2019. The future of food and agriculture | FAO | Food and Agriculture Organization of the United Nations. [ONLINE] Available at: http://www.fao.org/publications/fofa/en/. [Accessed 11 January, 2019].

2012). Increased competition for these resources could lead to their overexploitation and unsustainable use, degrading the environment and generating a destructive loop whereby resource degradation leads to perpetual competition for the remaining accessible resources, activating further degradation.

Land degradation, deforestation and water scarcities are among the most noticeable indicators of this unsustainable race:

- Worldwide, 33 % of the world's farmland is degraded either moderately or highly. This degradation distresses dryland areas and impedes the path to realizing food security and reducing hunger. Bringing that land into agricultural production would carry heavy environmental, social and economic costs (FAO, 2014).
- FAO estimates that more than 40 per cent of the world's rural population lives in river basins that are classified as water scarce (FAO,2011b).
- Globally, Agriculture is the proximate driver for about 80% of deforestation. Commercial agriculture is the most significant driver of deforestation in Latin America, accounting for around 66% of total deforested area. In Africa and tropical and sub-tropical Asia, subsistence agriculture accounts for a larger share of deforestation than commercial agriculture (Kissinger, Herold and De Sy, 2012).

Climate Change[4]

The Inter-Governmental Panel on Climate Change (IPCC) forecasts that temperatures in India are expected to rise by 3-4 degrees Celsius by the end of the 21st century (Pathak *et al.*, 2012). Combining these forecasts with estimates of Aravind Subramaniam *et al.*, 2018 (Climate Change and Indian Agriculture) imply that in the absence of any adaptation by farmers, such as change in cropping techniques or development in irrigation, agricultural incomes will fall by 12% on average, and by as much as 18% in unirrigated areas by the end of the century.

---

4 Ideas for India, 2019. Climate change and Indian agriculture. [ONLINE] Available at: https://www.ideasforindia.in/topics/agriculture/climate-change-and-indian-agriculture.html. [Accessed 9 January, 2019].

GHGs emission from Agriculture[5]

Agriculture's greenhouse gas emissions continue to rise; although not as fast as emissions from other human activities. Global emissions by sources from agriculture, forestry and other land-uses were more than 10 billion tonnes CO_2 eq in 2010. Global removals by sinks from agriculture, forestry and other land uses were more than 2 billion tonnes CO_2 eq in 2010.

The largest emitters in agriculture are Enteric fermentation (40%), Manure left on pasture (16%), Synthetic fertilizers (13%), Paddy rice (10%), Manure management (7%) and Burning of savannahs (5%). Livestock-related emissions from enteric fermentation and manure contributed nearly two-thirds of the total.

Food waste[6]

Fifty-four per cent of the world's food wastage transpires "upstream" during production, post-harvest handling and storage. Forty-six percent of it transpires "downstream," at the processing, distribution and consumption stages. As a general trend, developing countries suffer more food losses during agricultural production, while food waste at the retail and consumer level tends to be higher in middle- and high-income regions, where it accounts for 31-39 per cent of total wastage than in low-income regions (4-16 per cent).

Globally, around one-third of all food produced is lost or wasted along the food chain, from production to consumption (HLPE, 2014). Food losses and waste often translate into economic losses for farmers and others stakeholders within the food value chain, and higher prices for consumers, both of which affect food insecurity by making food less accessible for vulnerable groups. Reducing food losses and waste would increase the supply of available food and strengthen global food security.

---

5 UN FAO, 2014. Greenhouse Gas Emissions from Agriculture, Forestry and Other Land Use. [ONLINE] Available at: http://www.fao.org/resources/infographics/infographics-details/en/c/218650/. [Accessed 9 January, 2019].

6 UN FAO, 2013. Food waste harms climate, water, land and biodiversity. [ONLINE] Available at: http://www.fao.org/news/story/en/item/196220/icode/. [Accessed 9 January, 2019].

Projection and discussion

Agriculture 4.0 will no longer have to depend on applying water, fertilizers, and pesticides across entire fields. Instead, farmers will use the minimum quantities, or even completely remove them from the supply chain. They will be able to grow crops in arid areas and use abundant and clean resources, such as the sun and seawater to grow food.

Imagine a farmer who rather than depending on his intuitive characters, lets farm management software apply smart algorithms and a large pool of farm data to make the daily choices on how to work on his fields. Instead of sitting on machineries outdoors, the farmer with handful of associates can operate all the machineries for a farm covering ten and thousands of hectares from a room. GPS-guided autonomous UAVs shall constantly provide the data required for the algorithms, and other GPS-guided equipment works the farm with precision, every now and then even at the level of the individual plant. The crops have been genetically edited to resist most fungi, viruses, and insects and are highly efficient in their uptake and use of nutrients for the particular farm or the agro-ecological zone. As a result, the farmer needs less fertilizer, water, and crop protection than it did in the past.

This paper explores the key trends that will reshape food and agriculture in next two decades. Several megatrends are combing to intensifying the demand for agricultural outputs thereby promote the development of productivity enhancing innovations. Moreover, trends shaped by consumer preference will determine the specific methods, locations and outputs of agricultural production.

Vertical Farming and Urban Agriculture[7]

Agricultural activities within the city limitations have a long past in both, developed and developing countries. Particularly, in developing countries these activities contribute to food security and the mitigation of malnutrition. They generate additional income and contribute to recreation, environmental health as well as social interaction.

---

7 International Journal on Food System Dynamics, 2016. Urban AgriCulture and Food Systems Dynamics. [ONLINE] Available at: https://ageconsearch.umn.edu/bitstream/244536/2/46-A10_Hirsch-Terlau_-UrbanFoodSystem%20final.pdf. [Accessed 10 January, 2019].

People seek to change from traditional to new sustainable forms of food supply. From this evolves a demand for urban places that can be used agriculturally. Urban Farming is still a niche drive with the potential to contribute more significantly to urban development and establish a pillar of urban quality of life.

Hydroponics

Hydroponics is a method of growing plants without using soil (i.e., soil less). This technique instead uses a mineral nutrient solution in a water solvent, allowing the nutrient uptake process to be more efficient than when using soil[8].

US based startup Bright Farms[9] aims to help consumers buy local produce by building greenhouses in the communities it serves. So far, this strategy has resulted in leafy greens that take 80% less water, 90% less land and 95% less shipping fuel than lettuce that is grown out and shipped to Ohio.

Seawater Farming and Desert Agriculture

Most of the world's surface is covered in water, in the form of oceans. The remaining landmass of the Earth amounts to approximately 29 per cent of the surface. Of this remaining 29 per cent one-third consists of deserts of all types. To tackle the food crisis, the world must turn the world's desert and sea into food production facilities, an effort that will need the combined brainpower of the brightest minds, universities, and research facilities.

While many projects have demonstrated the feasibility of seawater farming in specific locations, little researches have been done to study seawater farming in a wide variety of conditions[10]. In addition, existing seawater projects rely on a combination of technology and indigenous socioeconomic factors, which might not be easily replicated elsewhere.

---

8 Maximumyield, 2019. Hydroponi cs. [ONLINE] Available at: https://www.maximumyield.com/definition/70/hydroponics. [Accessed 9 January, 2019]

9 Brightfarms, 2019. BrightFarms® | Our Mission | Locally Grown Produce. [ONLINE] Available at: https://www.brightfarms.com/our-mission/. [Accessed 11 January, 2019].

10 Seawater Farming | Mission, 2014: Feeding the World, 2019. Seawater Farming | Mission, 2014: Feeding the World. [ONLINE] Available at: http://12.000.scripts.mit.edu/mission, 2014/solutions/seawater-farming. [Accessed 10 January, 2019].

Existing pilot projects are running autonomously in Seawater Forest Initiative (Eritrea), Seawater Foundation, Bahia Kino (Mexico), and Seawater Green House (Australia).

Sundrop Farms in the South Australian desert manages to turns seawater and sunlight into energy and water in its greenhouse. Then use sustainably sourced carbon dioxide and nutrients to maximise the growth of crops[11].

Sustainable Packaging

Petroleum derived plastics dominate the food packaging industry even now-a-days. These materials have brought a lot of convenience and desirability to agro, food and packaging industry. These materials also have brought along with them problems relating to the safe-disposal and renewability of these materials. Due to the rising concern over environmental problems of these materials, interest has shifted towards the improvement and promoting the use of "bio-plastics". Bio-plastic is a term used for sustainable packaging materials derived from renewable resources i.e. produced from agro/food sources, materials such as starch, cellulose, etc. and which are considered safe to be used in food applications[12].

Bio-plastic could be the solution to the growing accumulation of hydrocarbon-derived plastics. With reports of growing amounts of plastic accumulating in the Earth's oceans, not to mention landfills, there is a real need for some form of biodegradable alternative.

The following bio-plastic solutions could substitute regular hydrocarbon derived plastics in the not too distant future. This list[13] is in no particular order and is far from complete.

---

11 Josh Hrala, 2019. This World-First Farm Grows Vegetables in The Desert With Nothing But Sun And Seawater. [ONLINE] Available at: https://www.sciencealert.com/this-farm-uses-sun-and-seawater-to-grow-vegetables-in-the-desert. [Accessed 10 January, 2019].

12 Starch Based Bio-Plastics: The Future of Sustainable Packaging, 2019. Starch Based Bio-Plastics: The Future of Sustainable Packaging . [ONLINE] Available at: https://www.scirp.org/Journal/PaperInformation.aspx?PaperID=84926. [Accessed 10 January, 2019].

13 13 Bio-plastic Innovations That Might Save the Environment, 2019. 13 Bio-plastic Innovations That Might Save the Environment. [ONLINE] Available at: https://interestingengineering.com/13-bioplastic-innovations-that-might-save-the-environment. [Accessed 10 January, 2019].

Bio-Plastic from Sugar and CO_2: Scientists from the Centre of Sustainable Chemical Technologies (CSCT) at the University of Bath managed to make a biodegradable form of plastic. It is also made sugar and carbon dioxide and not oil.

Bio-Plastic Containers can be Made from Rice: Finnish researchers, in 2014, were able to use rice starch to make sustainable, biodegradable polymers. These polymers could then be used to make food packaging.

Banana Peel as Bio-Plastic: Turkish startup by Elif Bilgin, managed to devise a chemical process for converting the peel into a resistant bio-plastic that would offer a solution to reduce waste and our dependency on oil for plastics.

Genetic Modification (Genetic Editing)

To nourish the escalating human population, it is vital that the world figures out techniques to increase food production. Increasing crop yields through conventional plant breeding is inefficient, the outcomes are unpredictable and it can take years to decades to create a new strain.

On the other hand, powerful genetically modified plant technologies can swiftly yield new plant varieties, but their acceptance has been debateable. Many consumers and countries have rejected GMO foods even though extensive studies have substantiated that, they are harmless to eat. But now, a new genome editing technology known as CRISPR may offer a good alternative.

Plant-Based Foods

Protein is one of the fundamental building blocks of the human diet, and powerful factors are shifting demand and driving the progress of novel sources. The existing supremacy of meat and seafood will fade in the coming decades, and quite a few alternative protein sources will fill the gap, potentially representing upto one-third of market share within the overall protein space by 2054[14].

---

14 Lux Research, 2019. Lux Research. [ONLINE] Available at: http://members.luxresearchinc.com/research/report/16091. [Accessed 10 January, 2019].

Growing 1 kg of kidney bean protein requires 18 times less land, 10 times less water, 9 times less fuel, 12 times less fertilizer, and 10 times less pesticides than producing 1kg of beef protein[15].

Plant-protein food choices should be encouraged globally to reduce the intolerable environmental footprint associated with the production of animal-protein foods.

Insect as Food

2.5 billion people around the world are already eating insects. That means there are more people dining on bugs than speak English! There are about 2,000 varieties of known edible insects, each with its own particular taste and texture. Introducing insects into food and cutting out less sustainable forms of protein, can help reducing many of the adverse impacts that food has on the environment. Clearance of rainforests for livestock rearing and soya production, the massive amounts of Greenhouse Gases emitted by the livestock industry, and the volumes of fresh water being sucked dry by an unsustainable food system.

Six Foods[16], a US based startup consider that six legs are better than four, and introduces insects to the Western diet as a sustainable food. Six Foods make chips with cricket flour and calls it as Chirps. They are 3x the protein and ½ the fat of the leading potato chip. Plus, they're non-GMO and gluten-free.

3D Food Printing Technology

People all over the world suffer from diseases, which require monitoring of their diet as part of treatment. The best possible option to solve this problem is to have a customized diet plan for each individual. 3-Dimensional (3D) food printing is an ideal technology which can lead to such mass customization of food. Apart from health benefits 3D food printing technology in food sector has a great potential

---

15 Cambridge Core, 2019. The environmental cost of protein food choices. [ONLINE] Available at: https://www.cambridge.org/core/journals/public-health-nutrition/article/div-classtitlethe-environmental-cost-of-protein-food-choicesdiv/DB40E5C12D662913CC342D3C19F85F7D/core-reader#. [Accessed 10 January, 2019].

16 Chrips, 2018. WHY CRICKETS?. [ONLINE] Available at: https://eatchirps.com/pages/why-crickets-1. [Accessed 10 January, 2019].

to fabricate 3D constructs with complex geometries, elaborated textures and tailored nutritional contents. Most research on 3D food printing is focused on improving the aesthetic quality of food in the short duration and public health in the long-term. The concept of 3D food printing represents a new forge in food processing and industry to manufacture new food products of variable shapes and colour and with particular composition[17].

Internet of Things

The farming and agricultural industry depends on innovative ideas and technological advancements to help increase yields and better allocation of resources. The late 19th century and the 20th century brought a number of mechanical innovations, like tractors and harvesters. Internet of Things (IoT) is the driving force behind increased agricultural production at a lower cost today, leaving the door wide open for the industry to bring a smart farming solution or IoT agricultural sensor to market.

It's projected that by 2024, over 225 million agricultural IoT devices[18] will be in use: The average farm will produce 4.1 million data points daily in 2050, up from 0.19 million in 2014. But while the growing number of connected devices epitomises a big prospect for agricultural and food industry, it also adds intricacy. The answer lies in making use of right technologies that help comprehend, reason, interact, inter-relate and increase efficiency.

Unmanned Aerial Vehicles (UAV)

Drones are not a new technology. But appreciations to investment and growing enabling regulatory environment, their time might have arrived. The value of drone-powered solutions in all applicable industries could be more than $127 billion[19]. Drone technology is giving agriculture a high-tech makeôver. Here are the few ways drones will be used throughout the crop cycle:

---

17 Acta Scientific Nutritional Health, 2017. 3D Food Printing: A Revolution in Food Technology. [ONLINE] Available at: https://actascientific.com/ASNH/pdf/ASNH-02-0050.pdf. [Accessed 9 January, 2019].

18 Brian Ray, 2019. An In-Depth Look At IoT In Agriculture & Smart Farming Solutions. [ONLINE] Available at: https://www.link-labs.com/blog/iot-agriculture. [Accessed 10 January, 2019].

19 PWC, 2016. Clarity from above. [ONLINE] Available at: https://www.pwc.pl/pl/pdf/clarity-from-above-pwc.pdf. [Accessed 9 January, 2019].

- Soil and Field analysis: By producing accurate 3-D maps for primary soil analysis, drones can play a role in scheduling seed planting and gathering data for managing irrigation and nitrogen levels.
- Planting: Biocarbon Engineering, a UK-based company backed by drone manufacturer Parrot, has come up with a method of planting trees quickly and cheaply. A drone loaded with germinated seeds fires pods into the ground at a rate of one per second[20].
- Crop spraying: Drones can scan the ground, spraying in real time for even coverage. Aerial spraying is up to 60 times[21] faster with drones than manual spraying.
- Crop monitoring & Irrigation: Multi-spectral imaging camera sensors on agricultural drones allow the farmer to manage crops, soil, fertilizing and irrigation more effectively.
- Plant Health Assessment: By scanning a crop using both, visible and near-infrared light, drone-carried devices can help track variations in plants and indicate their health and alert farmers.

In future, UAVs might involve fleets, or swarms, of autonomous drones that could tackle agricultural monitoring tasks collectively, as well as hybrid aerial-ground drone actors that could collect data and perform a variety of other tasks. However, safety of drone operations, privacy issues, and insurance-coverage still pose questions.

Blockchain and the Agriculture Supply Chain

Blockchain is a cryptic unalterable ledger for recording history of transactions. It is not centralized but distributed among every member of the blockchain. Blockchain technologies present an opportunity to transformationally improve the operations of agricultural supply chains, from farm to fork. Potential benefits abound for all supply chain actors, ranging from small-scale producers to middlemen (including processors, distributors and others) to end consumers.

---

20 World Economic Forum, 2017. These drones can plant 100,000 trees a day. [ONLINE] Available at: https://www.weforum.org/agenda/2017/06/drones-plant-100000-trees-a-day/. [Accessed 9 January, 2019].

21 Defence Web, 2018. Crop-spraying drones promise to change agriculture. [ONLINE] Available at: http://www.defenceweb.co.za/index.php?option=com_content&view=article&id=51771:crop-spraying-drones-promise-to-change-agriculture&catid=125:un. [Accessed 9 January, 2019].

Blockchain can reduce inefficiencies and fraud, and improve food safety, farmer pay, and transaction times. By improving traceability in supply chains, it can enable regulators to quickly identify the source of contaminated foods and determine the scope of affected products during contamination incidents. Additionally, the technology can reduce waste by detecting bottlenecks in the supply chain contributing to food spoilage.

In Arkansas, a small, farmer-owned agricultural co-operative called Grassroots, has partnered with Heifer International and the blockchain technology provider Provenance to create an interface that allows farmers and other supply chain actors to immutably track the path of poultry, as it moves through the supply chain.

Agri-digital is an Australian startup using blockchain in an analogous way in the Australian grains industry, helping to track and manage grains as they move throughout the value chain. Their platform also enables financial transactions in order to increase the speed by which farmers receive payment for their commodities, improving efficiencies across the whole supply chain[22].

Nano-technology in Agriculture

The Green Revolution of the 20th century was driven by a blind use of pesticides and chemical fertilizers, resulting in a loss of soil biodiversity and a rise in resistance against pathogens and pests.

Growing realization that conventional farming technologies would neither be able to increase productivity any further nor restore ecosystems damaged by existing technologies has paved the pathway to application of nano-technology in agriculture. Nano-technology has many applications in all stages of production, processing, storing, packaging and transport of agricultural products.

This technology will transform[23] agriculture and food industry by innovative new techniques, such as Precision farming techniques,

---

22 Next Billion, 2018. Blockchain for Agriculture: Improving Supply Chain Efficiency and Access to Finance for Smallholder Farmers. [ONLINE] Available at: https://nextbillion.net/blockchain-for-agriculture/. [Accessed 9 January, 2019].

23 International Journal of Chemical Studies, 2018. Application of nano-technology in precision farming: A review. [ONLINE] Available at: http://www.chemijournal.com/archives/2018/vol6issue5/PartN/6-4-773-173.pdf. [Accessed 9 January, 2019].

controlled release of agro-chemicals and site targetted delivery of various macromolecules needed for improved plant disease and pest resistance, efficient nutrient utilization and enhanced plant growth. Processes, such as nano-encapsulation show the benefit of more efficient use and safer handling of pesticides with less exposure to the environment that guarantees eco-protection.

Crowd farming

Crowd farming is a model where a number of investors share risks and own shares in a farming venture. This allows those that have work commitments away from the land to participate and invest in farming projects on a large scale. The origins of the 'crowd' concept can be traced to crowd-sourcing where a large number of people are invited to participate in a project[24].

Social entrepreneurial project, Naranjas del Carmen, has developed the concept of Crowd farming. Naranjas del Carmen has created a system in which the person has ownership over the trees and land that the farmer cultivates. In this way, the fruit of those trees goes to their owners, creating a direct link between production and consumption and avoiding overproduction and waste along the value chain.

Conclusion

Altogether these new technologies are shifting the way stakeholders think about the agriculture industry, giving hope to solving the hunger and food scarcity problem. This transformation offers a great opportunity for government to take a leading position in the Agriculture 4.0 era.

Although forecasts can shed light on the future, we are still 32 years away from 2050. A complete new generation of farmers, who are not yet born, will be farming mid-century, and much will materialise between now and then that we cannot envisage. But if the past is a hint to the future, then the growers will remain to seek improved ways to harvest crops by embracing innovation.

---

24 New Zimbabwe, 2017. Is crowd farming an option for Zimbabwe?. [ONLINE] Available at: https://www.newzimbabwe.com/is-crowd-farming-an-option-for-zimbabwe/. [Accessed 9 January, 2019].

References

BCG, 2018. Crop Farming, 2030. [ONLINE] Available at: https://www.bcg.com/en-in/publications/2015/crop-farming-2030-reinvention-sector.aspx. [Accessed 11 January, 2019].

The future of food and agriculture | FAO | United Nations, 2019. [ONLINE] Available at: http://www.fao.org/publications/fofa/en /. [Accessed 11 January, 2019].

Maggie McGrath, 2019. The 25 Most Innovative Ag-Tech Startups In 2018. [ONLINE] Available at: https://www.forbes.com/sites/maggiemcgrath/2018/06/27/the-25-most-innovative-agtech-startups-in-2018/#6768aa262302 . [Accessed 11 January, 2019].

OUP Academic, 2019. Agriculture in 2050: Recalibrating Targets for Sustainable Intensification | BioScience | Oxford Academic. [ONLINE] Available at: https://academic.oup.com/bioscience/article/67/4/386/3016049 . [Accessed 11 January, 2019].

AgFunderNews, 2019. Accelerators & startups Archives - AgFunderNews. [ONLINE] Available at: https://agfundernews.com/category/industry/accelerators-startups . [Accessed 11 January, 2019].

❑❑❑

Chapter **10**

Role of Agri-business Incubation Centres of Research Institutions in Promoting Startups

Sudha Mysore

Principal Scientist & In-charge Agri Business Incubation, ICAR-IIHR, Bengaluru
E-mail: sudham@iihr.res.in

Innovation and Entrepreneurship, in recent times, have emerged as key development goals for attaining the twin objectives of societal welfare with economic growth. Innovation packaged as improved technical knowledge has globally been accepted as the basic tenet of the new age development paradigm. Research organizations/ universities world over have been the repositories of such knowledge generated over decades of dedicated research efforts (Sampat, 2009). Effective and widespread dissemination and adoption of such knowledge is the key component of this growth trajectory, especially so in agriculture. Since Research Organizations / Universities hold the primary mandate of continued innovation, a separate dedicated system is mandatory for the continued transfer of technical knowledge or know-how to benefit the end-users along the technology value chain. Such a system needs to be responsible not only for supply of technology produce at appropriate time, at affordable cost but also ensure its supply in adequate numbers. Though the Research Organizations and Universities encouraged their own extension services for the dissemination and technology transfers, their role gets limited to extending service and not provide solution for commercial scale production of technology products. The state run services also are unable to bridge this gap, there by leading to the emergence of a middle agency that helps research organizations and universities meet the growing demand of their technology products, popular as Technology Transfer Offices (Sudha, 2015).

The concept of Technology Transfer Offices (TTOs) as incubation facilities that promulgate entrepreneurship development for technology commercialization is well-adopted and accepted across developed countries. A plethora of such efforts, models and success stories of innovation lead societal welfare have all been documented world over (Jaffe, 1998; Cohen, Florida, Randazzese, and Walsh, 1998). The enactment of the Bayh-Dole Act, 1980 in the US is often considered a milestone that changed the university-industry linkage in technology transfer and commercialization. Evidence from the US universities and research organizations suggests strengthening of university industry linkage post Bay-Dole Act with technology transfer and commercialization efforts generated a four-fold increase in numbers of licenses and royalty incomes [Arlington, VA 2006], suggesting that innovation lead incubation and technology transfer directly or indirectly encouraged Public Private Partnership (PPP) as well.

Emerging economies like Brazil and Chile also have successfully emulated the concept of TTOs and commercialization through incubation aiming at creation of knowledge societies in the long run. In view of the changing global order, falling in line with other emerging economies, India also has adopted similar policy of initiating innovation centres and TTOs. This function of technological innovation in the development paradigm is of particular relevance to the agricultural research of countries like India, given their mandate of crop productivity enhancement linked to societal welfare.

The Union Government of India also has been promulgating the policy of encouraging startups and innovation focused programmes through StartUp India and Stand Up India missions. The R&D institutions can play an important role in this endeavour through technology support and facilitating the promotion of startups.

The Indian Council of Agricultural Research (ICAR), the apex body responsible for all agricultural research and development for India is the only organization that appears to have not only adopted its own Intellectual Property Rights (IPR) policy in 2006, but has successfully implemented it across its 90 and odd research institutions by setting up Technology commercialization and incubation offices since 2009 with the funding support from the World Bank aided National Agricultural Innovation Project (NAIP). Nearly over 60 odd institutions under ICAR have been operating successful TTO and Incubation facilities, the Indian Institute of Horticulture Research(

ICAR-IIHR), Bengaluru being one of them having a successful incubation facility since 2013.

These BPDs colloquially known as 'incubation centres' are anticipated to serve as knowledge centres that groom a budding enterprise into a successful model ready to face the market. The competitiveness to serve the market better and converging co-operation with business roles and rules are vital links for the success of a new enterprise. Often the absence and or negligence of these links prove to be detrimental and self-destructive, leading to a vicious circle. BPDs ensure these positive and productive engagements by tailoring support for their incubatee companies and establishing links with critical resource networks like seed capital providers, investors and professional business experts.

In this direction, the incubation and business planning and development models adopted by different institutes under ICAR have resulted in initiation and development of agri- entrepreneurship across regions. This paper discusses in detail business entrepreneurship development approaches adopted by Horticultural Technology Management and Business Planning and Development (HTM-BPD) unit in the field of horticulture keeping in view client requirements and highlights the role played by institutions like ICAR-IIHR through the incubation facility in achieving regional development goals and enhancing the institutional visibility through technology/knowledge transfer.

Horticultural Technology Management & Business Planning and Development Unit at ICAR-IIHR

A total of 26 research and development institutions under the aegis of ICAR are the primary sources of technical support for the horticulture crop production and the associated industry. Indian Institute of Horticulture Research (ICAR-IIHR), a premier institute has developed and disseminated a large number of improved varieties, hybrids, and provided the package of practices for the cultivation of over 40 horticultural crops. With the growing demand for horticultural crop varieties, technologies and associated inputs, ICAR-IIHR in the recent past has adopted a new Intellectual Property Rights (IPRs) policy and enhanced its visibility through technology commercialization through licensing. Falling in line with the ICAR's IPR policy, the ICAR-IIHR has initiated its technology transfer and commercialization efforts through its Institute Technology Management Unit working since 2008. The underlying objective of this policy being ensuring adequate supply of quality seeds and planting material and all other inputs including plant health management techniques, post-harvest technologies through authorized licensees for wider spread and adoption. ICAR-IIHR has been successful in commercializing over 35 technologies and harnessed more than Rs. 9 crores gross realization within the last decade. The technologies include seed and planting material, bio-pesticides, balanced micro-nutrient foliar formulations, and pest and disease management tools. The institute licenses its technologies to both, private and public sector organizations, individual entrepreneurs and producer organizations on non-exclusive basis.

In view of the growing significance of horticulture and the potential for technology commercialization, it became imperative that these technology licensees also be given further technical support and handholding for a period of six months to one year through incubation. ICAR-IIHR's focus, thus, started changing towards re-defining the technology commercialization from a mere information/technical support to infrastructural & training support that is inclusive of technology business process & market information research that would equip the entrepreneur with the requisite technical and market intelligence for scaling up the technical innovations into business proposals through handholding and mentoring through incubation. ICAR-IIHR's HTM BPD is, thus, an innovative step forward to equip producers into entrepreneurs and entrepreneurs into business leaders with the specific objectives of :

- Technology commercialisation through incubation for horticultural production systems
- Enterprise and Entrepreneurship development in horticultural production systems and
- Capacity building and imparting consultancy skills for horticultural crop-based entrepreneurs.

While the demand for different groups/categories of technology products differs, the demand for plant health management technologies inclusive of bio-pesticides and crop specific micro-nutrient foliar spray formulations has been rising over the years. Nearly 353 companies representing both, public and private sectors and varying scale of operation are ICAR-IIHR's clients. The gross revenue generated from technology licensing has been increasing year after year, fetching the institute steady flow of income/IP lead revenue.

Models of Incubation at Horticultural Technology Management Business Process Development Facility

The Business incubation facility of ICAR-IIHR provides technology commercialization and incubation under three distinct models, i.e., On-site incubation, Off-site incubation and specialized incubation as explained below:

On-Site Incubation: Technology Commercialization through Value Chain Incubation

As a technology-based incubation facility, Agri-Business Incubation (ABI) facility provides its member clients, an opportunity to scale up the licensed technology for maximizing profits. Member Clients, who are selected after rigorous criteria, are provided with technology up-scaling facilities inclusive of infrastructure, space and handholding support by technical experts. The Scientist innovator works in close collaboration with the client for scaling up the processes. ABI also permits the client to sell their produce through the marketing network of the research institution and assist the client get better market. Besides the technology licensing fee, the client pays additional pre-fixed rent for the facilities provided and also a royalty in the range of 5-10% on the net sale of the produce sold through the institutes' sales network. An On site incubatee is permitted space and facilities for a maximum of 24 months before the client graduates from the incubation process. Some of the services that are included for the on-site incubation are, Business Development Services: Navigating regulatory Requirements, Standards and Compliance. Product / Process/Human Resources Development & Management; Facilitate Marketing Assistance: Promotion through Trade Fairs, Exhibition and Industry Meets; Facilitate Finance: Bankable proposal / Angel Investors, Finance Management Skills, etc.

Off–Site Incubation: Technology Commercialization

The off–site incubation pathway differs from the former one, only in the fact that the client is not operating from the institutes premises. The client does license a technology from the institute and is provided services similar to those of on-site incubation on payment, except for the permission to market their produce through institute's sales counters. The TTO's officers in collaboration with the scientist innovators visit the client's site and provide the requisite advice and direction for up-scaling the licensed technology.

Specialized Incubation

The third incubation path way is a value proposition incubation, where in the innovators provide consultancy services as per the specific requirements of the clients and or an opportunity for research collaboration with the institute. The facilities include Research Collaboration: Concept Testing, New Product Development, Product Enhancement; Paid up Trials: New Product Testing; Consultancy: Farming, Need-based Consultancy and IPR Services: Patent Search, Patent filling, etc.

Performance and Success Stories

ABI at ICAR-IIHR, which has started operations during May 2013, has successfully initiated the processes of incubation and business process development and has so far supported around 7 on-site incubatees successfully, while more than 100 have benefitted from the technical guidance and support provided by the incubation facility.

Besides providing incubation facility, the ABI also focuses on organizing Entrepreneurship Development Programmes (EDPs), business mentoring and awareness campaigns for enthusiastic horti-prenuers and research scientists, facilitate research collaborations, patent search facilitation and patent filing.

Impact, Lessons Learnt and Way Forward

ABIs, colloquially termed 'incubation centers' have been visualized as potent components that foster systematic and comprehensive Agri-entrepreneurship development by linking innovations with enterprise development to strengthen agri-business entrepreneurship development and commercialization of technologies through incubation. As repositories of knowledge, incubators evolve as the social memory of risk taking venture processes and map the learning for better enterprise development continuously, day in and day out. It is important that these unique experiences of Incubators in knowledge-based enterprise development for creating new innovation society, are analyzed, understood, disseminated and practiced with suitable adaptation at incubators level. This would improve our delivery capability, greatly as a performing incubation community and it is needless to add the immense benefits, it can bring in at the enterprise level for the budding entrepreneurs.

The effort put forth by research institutions in technology development is said to be complete only, when it is adopted by the end-user, the ultimate consumer, which is possible only if the technical knowledge is transformed into a product suitable for commercial production and application. It is, thus, imperative that the university /research organizations understand the needs of the industry or work in close collaboration with the industry. The research institutions being primarily repositories of technical knowledge, may not be well equipped to meet the demand of their products or processes in adequate numbers as per the demand. Having an effective partnership with the industry or building, a franchisee will help them scale up and mass produce their products to benefit the end-users. Technology licensing by the research institutions for commercial licensees is one step in this direction, while creating facilities for hand-holding and up-scaling within or outside their own premises is several steps forward in technology transfer efforts. The year long journey of HTM BPD of ICAR-IIHR, in this direction has resulted in a number of positives and few lessons as well.

Impact

The presence of ABI at research institutions have attracted a number of enthusiastic budding entrepreneurs, state horticultural department officials, bankers and other stakeholders towards horticultural crop technologies and Horti-business enterprises alike. Through the on-site and off-site incubation facilitations, seven entrepreneurs have successfully operationalized their business ventures, more than 1000 jobs have been created as a direct impact and businesses worth crores have been established. The visibility and acceptability of the institute has nearly doubled with the incubation facility. State horticultural departments are approaching for closer collaborations and technology solutions from the institute.

Lessons learnt

- Horticultural technologies and value added products besides production of horticultural crops offer huge potential and offer scope for Horti-entrepreneurship development. Providing hands on training and handholding through scaling up process is essential to help start successful enterprises.

- While imparting technical Know-how or improved technology is important, equally important is to hold one-on one mentoring on business development and positioning.
- Creating holistic awareness programmes and linking with all the stakeholders along the value chain are important for successful scaling up.
- Linking the startups with angel financiers, venture capitalists and other sources is an essential feature of successful business development process.
- The technology licensing and commercialization is a two-way learning process, where in the innovator and the entrepreneur need to understand the ethics of 'licensing and commercialization'. The non-exclusive license process needs to be understood thoroughly by the licensee. At the same time, innovator needs to learn to understand the expectations of the licensee. The processes involved in scaling up, market capture and sustainability are the domain of the licensee alone.
- A technology, which has already been licensed for commercial production, should not be demonstrated by the innovator, unless the technology transfer clearly demarcates the zone of operations.
- There is an urgent need for the research institutions and the council to periodically evaluate the processes and make necessary amends in the policy to make these efforts sustainable in the long run.

References

Robbins, C., 2006. *Measuring payments for the supply and use of intellectual property*, Washington, D.C., USA: Bureau of Economic Analysis, U.S. Department of Commerce.

Sampat, B.N., 2009. *The Bayh-Dole Model in Developing Countries: Reflections on the Indian Bill on Publicly Funded Intellectual Property*, Washington, D.C., USA: United Nations Conference on Trade and Development and International Centre for Trade and Sustainable Development.

Sudha Mysore, 2015, Technology Commercialization through Licensing: Experiences and Lessons-A Case Study from Indian Horticulture Sector, *Journal of Intellectual Property Rights* Vol. 20, November 2015, pp. 363-374.

Slaughter, S. & Rhoades, G., 1996. The emergence of a competitiveness research-and-development policy coalition and the commercialization of academic science and technology. *Science Technology and Human Values,* 21(3): 303–339.

The World Intellectual Property Organization, 2012. *The World Intellectual Property Organization.* [Online] Available at: http://www.wipo.int/ipstats/en/statistics/country_profile/countries/br.html

World Bank, 2009. *Chile: Fostering Technology Transfer and Commercialisation,* Washington, D.C., USA: Latin American Region, Poverty Reduction and Economic Management Unit.

Wu, W., 2010. Managing and incentivizing research commercialization in Chinese Universities. *The Journal of Technology Transfer,* 35(2): 203-224.

Zuniga, P., 2011. *The State of Patenting at Research Institutions in Developing Countries,* Geneva, Switzerland: World Intellectual Property Organization.

❑❑❑

Chapter **11**

Opportunities for Agri Startups in Genomics and Plant Molecular Biology

Santosh R.S.[1], Siby Varghese[2], Shelba Mariam Skariah[1] and Basil Kuriokose[1]

[1] Phytocom Pharmaceuticals Private Limited, Kochi-683503, Kerala
[2] ESAF, Mannuthy-680651, Kerala, E-mail: santhosh@phytocom.co.in

Introduction

Plants support human life by mixing water and CO_2 using sunlight. Food, medicines, clothes, wood, fuel, papers, industrial raw materials and many tools are made available from plant woods. Agriculture is the domestication of plant for food, started 12,000 years ago as a result of human settlement. As population increased there was increased demand for food and led to intensive agriculture. To increase production more land was brought under cultivation, when land became scarce high yielding varieties of plant were made using human intelligence. Green revolution is a result of selection of best genetic breeds of agricultural crops. Next revolution in agriculture is expecting from evolving agri-startups, which are formed by scientists, engineers and industrialists using new technologies like CRISPR-Cas 9 induced genome editing. Agri-startups are equipped with new knowledge which can replace original plant breeding, protection or cultivation techniques. It take 20 years to make a hornless cattle through conventional breeding, which is achievable through genomic technologies in a year. Personalised, regionalised, stress tolerant, disease and pest resistant crops in a very less time become possible by them.

Recombinant Technology

A desired traits from any organism to another is possible through recombinant DNA technology, which enable researchers to transform original plants and animals. The technology was evolved in the 1970s and the first product in agriculture, Flavr-savr tomato, received a license for human consumption by FDA only in 1994. At present, genetically modified crops are available in several species including apples, barley, and potatoes. About 90 per cent of the corn, cotton, and soyabeans planted in the United States are genetically modified. By the end of 2020, more than 18 million farmers in 26 countries may adopt genetically modified crops. Transgenic maize plants overexpressing the *BcWRKY1* gene were generated by *Agrobacterium* mediated transformation, which were able to resist 300mM NaCl stress pending approval for cultivation. Even though there is less adoption of modern genetic technologies into agriculture, few genomic companies are working on it, produce newer agricultural plants with more hybrid vigour. There was increase in production through genetic technologies from 1996 to 2013 for soyabean and maize as 110 and 195 million tonnes, respectively.

Genomics

National Plant Genome Initiative is a dedicated step for advancing crop improvement through improving crop genome. Post-sequencing analysis will help to improve crop yield by innovating newer genetic traits. Agri-startups, which can go globally can overcome the stringent regulations for genetic test approvals and the high cost involved in testing. Recently agri-genomics are getting more importance, which is a branch of science dealing with the study of genetic makeup and their role in the development of crops and plants by completely sequencing the whole chromosomes. Agri-genomics is segmented on the basis of sequencing method, services, and application. The application of gene expression data and next-generation sequencing techniques plays a vital role in agri-genomics. The plant evolution, product yield, disease resistance, pest management etc can be studied.

Agri-genomics market is growing at a significant CAGR owing to increase in the government initiations for improving crop yield. In 2015, Genestack received £90,000 grant from Innovate UK to collaborate with Rothamsted Research on the development of agri-genomics for the development of crops. In 2016, Affimetrix, Inc. expanded its agri-genomics genotyping portfolio with new targetted sequencing solution. A**gri-genomics market** has a prominent growth in USA, which is attributed to increase in the preference for improving crop yield, rise in R&D activities in agri-genomics, and government support to undertake advanced sequencing techniques. But in **Europe**

agri-genomics market is driven by adoption of novel technologies, collaboration between the companies for development of newer techniques, and increase in agriculture production. **Asia Pacific agri-genomics market** growth is attributed to increase in government investment in agri-genomics. In India, China and Japan, collaboration between research institutes and industry are determining the market. Oryza sativa, two strains of rice (indica and japonica) became the first crop plant whose genome was sequenced and released in 2002, starting point in agri-genomics.

Cloud Biology

Temporal transcriptional data can reveal how genomes drive genes for crop production. Sequencing many closely related species predict the most deleterious alleles, and using machine learning programme, one can model transcriptional factor binding preferences and predict allele-specific expression within and across species. The Functional Annotation of Animal Genomes (FAANG) Consortium is dedicated to improving genome annotation for domesticated animal species worldwide. The allelic databases are available as Cotton Marker Database; VegMarks and Pea-nut Marker Database, Foxtail millet Marker Database (FmMDb). For addressing the importance of transcription factors, many databases have been developed like Plant Transcription Factor Database (PlantTFDB), Database of Arabidopsis Transcription Factors (DATF), Database of Rice Transcription Factors (DRTF), Database of Poplar Transcription Factors (DPTF). Two million SNPs and INDELs that would provide informational resources to help further marker-assisted selection in rice breeding programs. The first generation molecular marker technology like Restricted Fragment Length Polymorphisms (RFLP), Random Amplified Polymorphic DNA (RAPD), Single Nucleotide Polymorphism (SNP), Amplified Fragment Length Polymorphism, (AFLP) and Sequence Related Amplified Polymorphism (SRAP), etc. are still valid compared with the traditional morphological markers, cytological markers, and biochemistry markers. DNA markers are widely used in identification of crop varieties, conservation and evaluation of plant germplasm resource, analysis of genetic adversity and evolution, construction of genetic maps, cloning of important agronomic trait genes, and molecular mark assisted breeding. DNA molecular marker has high specificity, selectivity, simplicity, accuracy and genetic stability can reflects the differences in DNA level without environmental impacts, and thus, has great advantages in seed purity identification. It has great theoretical and practical significance in fields, such as Quantitative Trait Loci (QTL) mapping, marker assisted selection, gene map-based cloning and comparative genomics, etc.

Comparative Genomics

Comparative Genomics helps in understanding genomics information across tissues, developmental stages, natural diversity, and to define targets for crop improvement by advanced breeding or precision engineering. iDEP is such user-friendly web application for exploratory data analysis, differential expression, and pathway analysis of RNA-seq data. To improve genetic quality, many proprietary technologies are owned by the private sector.

Construction of the genetic maps, detection and isolation of genes responsible for the agricultural traits were the plant molecular science. Genetic maps described genomic regions associated with many phenotypic characters. Many different reverse genetic methods, such as transcriptomics, proteomics, RNAi gene silencing are used for detection of genes function, elucidation of complex biological events and specifying related genes. There are reports which describe molecular mechanism of cell structure, functions and signalling, growth and development, photoperiod and flowering, abiotic stress tolerance, grain yield and quality, disease resistance, etc. Today, many plant species including the enormous genome of bread wheat have been fully sequenced. Earlier plant genetics scientists who were concentrating on single gene effects are able to describe plant phenotypes at multi-genic effects. The whole genome sequencing produced high density genome-wide association maps that is a powerful tool for investigating the genetic relations between crops. Comparative genomic analysis of different agricultural crops are accelerating the discovery and utilization of valuable genes and alleles.

Construction of Synthetic Genome

Because of advancement in genomics many synthetic genome for plant can be designed. With the help of new gene technologies, large insertions and deletions of gene complexes can be done to make synthetic genome in agriculture. Plants with minimum number of required genes and with a gene cluster that allow the cells to divert metabolic resources for making only useful products can be made.

Smart plants which can sense change in environmental conditions thrive better in adverse climatic conditions. Sensing the soil conditions plants can do nitrogen fixation or mineralisation. From an agronomic point of view, genetic manipulation or metabolic channelling have led to altered emission of aroma profile from volatile production in crop plants, which may lead to desirable changes in plant defence, pollinator attraction and post-harvest qualities. New codon usage to incorporate unconventional amino acids will help the propagation of microbes through insect pests. The potential to increase the capacity and efficiency of plants to fix atmospheric carbon has clear implications for agricultural productivity and natural resource management. By cloning genes responsible for use of non-photosynthetic CO_2 capture pathways. This will prevent accumulation of CO_2 in greenhouses.

Pharmacogenetics in Agriculture

Agriculture includes livestock and poultry, gene manipulations to increase quality of meat and egg are possible. There are parallels between the pharmaceutical and agricultural chemical industry. Personalised medicine is a common talk in human diseases, genomics based drugs are now under development for plants. Genome based drug in agriculture was developed by Monsanto for application of an herbicide, Roundup, the crop was made herbicide resistant by cloning enzyme, which can metabolize the herbicide in plant. Better understanding of the pharmacogenetics variation across the plant kingdom is necessary to design an herbicide or pesticide. High-content screening platforms are required to develop new herbicides and agricultural chemicals with less cost input. How to invest a reasonable input of nutrient according to genome, at the same time guarantee the quality and yield of crops, and how to choose the more effective fertilizers according to plant genome. Development of growth regulators enhance useful plant attributes for yield increase. The green revolution was driven by breeding semi-dwarfed varieties for decreased gibberellic acid (GA) bio-synthesis in rice and GA signalling in wheat. Plant hormones, such as auxin, GA, and ABA continue to be excellent targets for the Ag-chemical industry.

Metabolic Engineering

Metabolic engineering of plants is another aspect. The development of Golden Rice, engineered to synthesize β-carotene in its seeds demonstrated the potential of metabolic engineering to enhance nutritional value of staple crops. Consuming â-carotene, the precursor to Vitamin A, can help combat malnutrition in some of the world's poorest populations. *Trichoderma hamatum* strain GD12 is unique in that it can promote plant growth, provide biocontrol against pre- and post-emergence soil pathogens and can induce systemic resistance to foliar pathogens. The genomic traits controlling this function in *T. hamatum* can be transformed into other strains. Bio-remediation of land using engineered microbes to alter the soil conditions so that agriculture crops can be grown in non-arable land.

Nutrigenomics

Nutrigenomics (or 'nutritional genomics') focuses on understanding how diet affects gene expression. Nutrigenomics relates the food consumed and its effect on gene expression in a cell, integrating high through put transcriptome, proteome and metabolome profiling techniques. Application of nutrigenomics is mainly in livestock and poultry. Nutrigenomics is also used in aquaculture to investigate the effect of diet upon production and response to stress. Recently, nutrigenomics identified factors that affect milk fat synthesis in ruminants. It is expected that the outcomes of nutrigenomics studies will impact how animal health and productivity is managed, in part by designing species and condition-specific diets for production animals.

Transcriptomics

Using transcriptomics techniques, researchers are discovering candidate genes to guide the design of targeted breeding strategies for improving milk production, meat quality in farm animals, egg production in chickens and understanding the pathogenesis of complex diseases. Micro-arrays, both commercially and non-commercially produced, are readily available for many livestock and aquaculture species. Frequent milking increases milk yield in dairy cows; micro-array analysis of dairy cows with different milking regimes identified 29 genes involved in persistent milk yield. These genes can be used as the basis for a bio-marker test for increased lactation times in dairy herds, similarly for meat quality. Bio-markers, such as circulating nucleic acids can now be used successfully in pregnancy diagnosis in cattle. Likewise, Reproductive efficiency in pigs were determined by understanding gene expression at early stages of development.

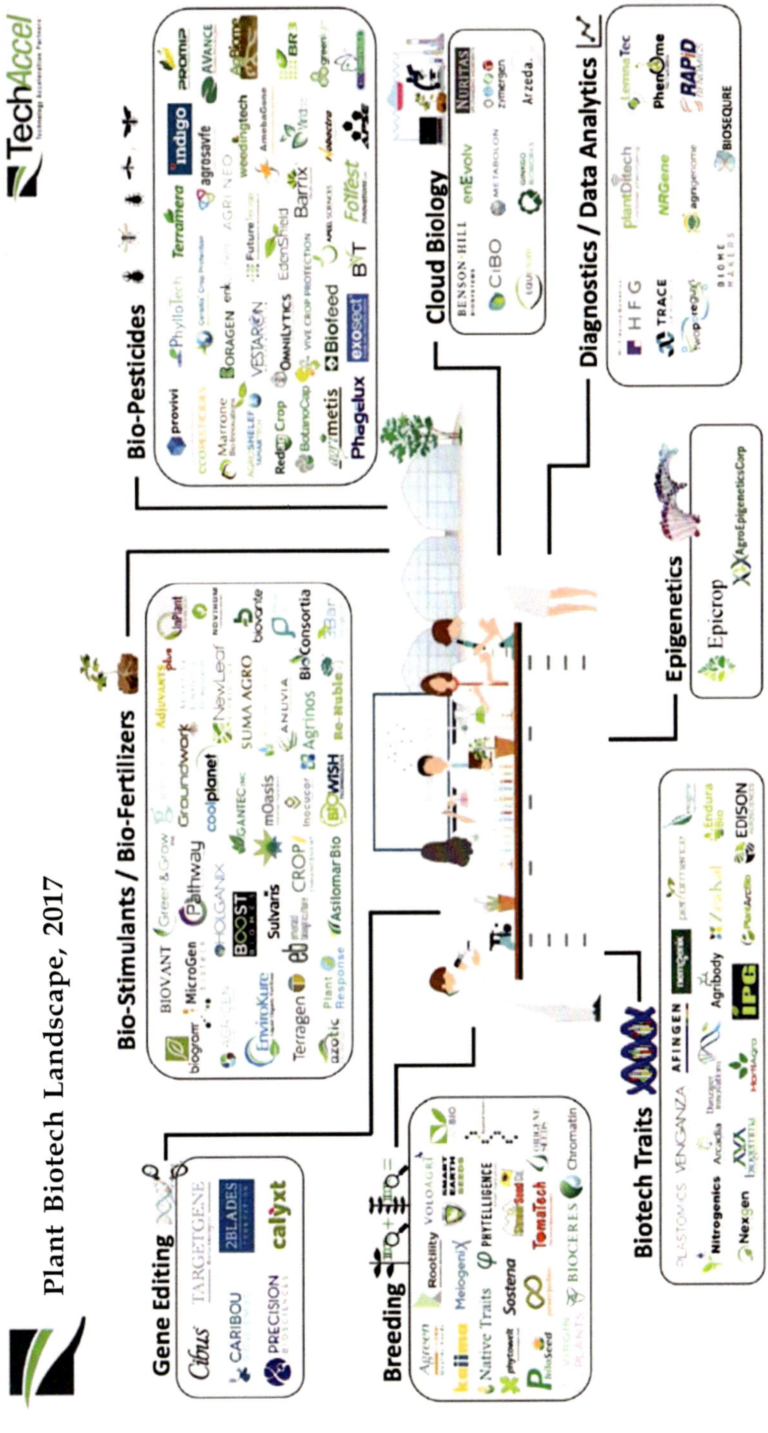
Plant Biotech Landscape, 2017
TechAccel
Gene Editing
Breeding
Bio-Stimulants / Bio-Fertilizers
Bio-Pesticides
Cloud Biology
Diagnostics / Data Analytics
Epigenetics
Biotech Traits

Proteomics

Proteins undergo a wide range of post-translational modifications including phosphorylation, ubiquitination, methylation, acetylation, glycosylation, oxidation and nitrosylation. Proteomics is more commonly used to investigate systems where it is necessary to directly identify proteins, for example, milk composition, gamete contribution to fertility and direct identification of immunogenic microbial proteins. Proteomics is also used to look at the outcomes of breed differences, even if the animals are closely related genetically. These proteins can serve as bio-markers for characterization of cattle breeds. Many individuals are allergic to egg proteins, genes responsible for that can be deleted from chicken genome to produce eggs not having allergic proteins.

Metabolomics

Metabolomics is the systematic study of metabolites (the metabolome) in a cell, tissue, organ or body fluids. It can be used in aquaculture, for estimating milk and meat quality attributes and in poultry. It is used for estimating breed-specific traits, such as fat content in cow milk, meat quality in pig by estimating the protein levels, fat accumulation and behaviour of muscles after slaughter. In poultry industry, discover new chemical contamination in chicken eggs.

Model Plants to Study Plant Molecular Biology

Physcomitrella patens.a bryophyte is gaining popularity among scientists for functional analysis experiments. Owing to multitude of properties, it possesses like small genome size (just thrice the size of *Arabidopsis* genome), dominant haploid gametophytic stage and its ability to integrate transforming DNA at high frequency by way of homologous recombination, *Physcomitrella* is gaining popularity as the new model plant for functional analysis.

Epigenetics

Genome-wide methylation mapping helps to understand epigenetics and used to study chicken, sheep and bovine embryonic development. It helps in understanding how gene expression patterns are inherited and how environmental changes contribute to regulation of gene expression.

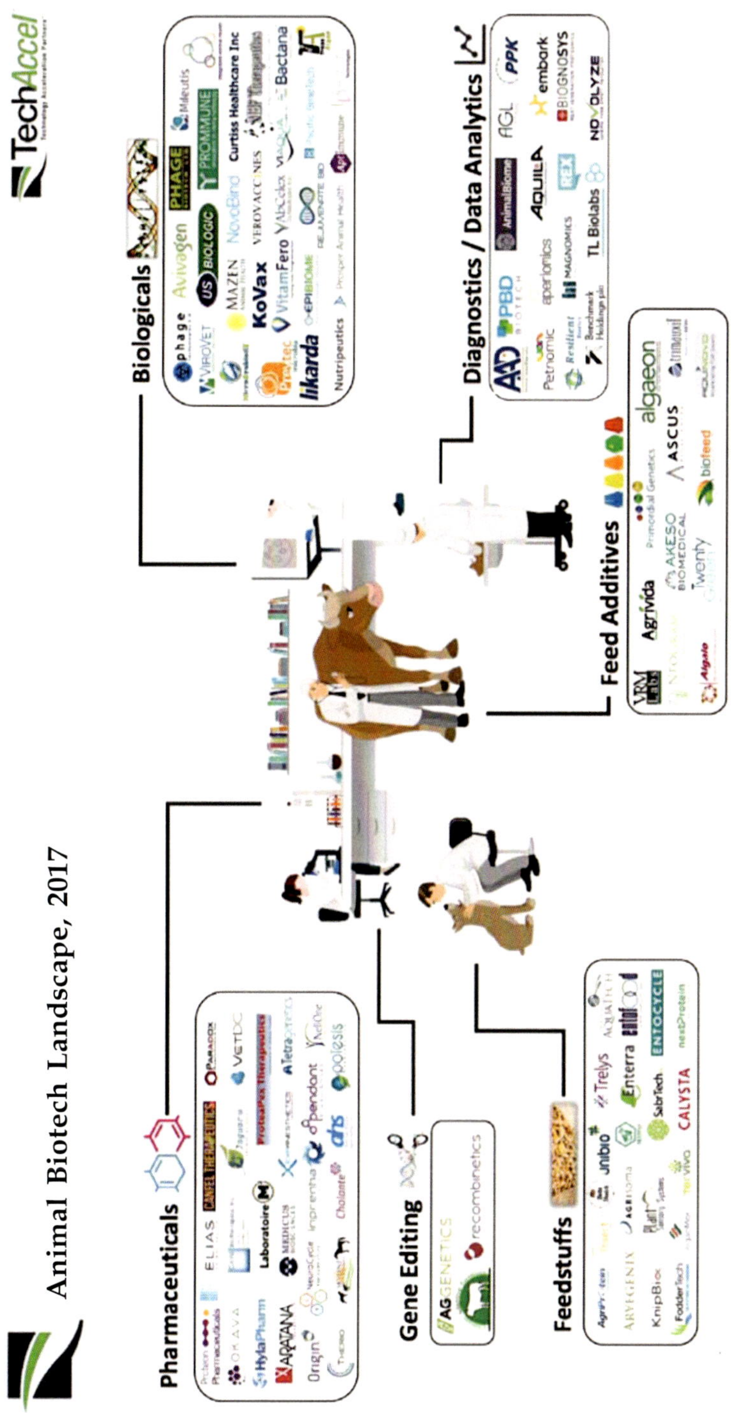
Animal Biotech Landscape, 2017
TechAccel
Pharmaceuticals
Gene Editing
Feedstuffs
Biologicals
Diagnostics / Data Analytics
Feed Additives

Conclusion

The commercialization of genetically modified crops dramatically lowering the cost of agriculture, and bring continuously a great deal of benefits of economic, environmental and social in the world. Transgenic technology is widely used to control the agricultural disease and insect-pest, reduce the cost of agricultural production, improve the effect of disease and insect control, reduce pesticide pollution, and increase crop production. Modern DNA technology becomes more and more widely used in agriculture, the current major issues like food, population, pollution and other problems that our world facing would get better solution and agri-startups can do a good role.

References

Beiki H., Eveland A.L., Tuggle C.K., 2018. Recent advances in plant and animal genomics are taking agriculture to new heights. Genome Biol 19: 48.

Di Berardino M.A., 2001. Animal cloning—the route to new genomics in agriculture and medicine. Differentiation 68: 67-83.

Fischer K.S., Barton J., Khush G.S., Leung H., Cantrell R., 2000. Genomics and agriculture. Collaborations in rice. Science 290: 279-280.

Harper K.N., Armelagos G.J., 2013. Genomics, the origins of agriculture, and our changing microbe-scape: time to revisit some old tales and tell some new ones. Am J Phys Anthropol 152 Suppl 57: 135-152.

Harper L., Campbell J., Cannon E.K.S., Jung S., Poelchau M., Walls R., Andorf C., Arnaud E., Berardini T.Z., Birkett C., Cannon S., Carson J., Condon B., Cooper L., Dunn N., Elsik C.G., Farmer A., Ficklin S.P., Grant D., Grau E., Herndon N., Hu Z.L., Humann J., Jaiswal P., Jonquet C., Laporte M.A., Larmande P., Lazo G., McCarthy F., Menda N., Mungall C.J., Munoz-Torres M.C., Naithani S., Nelson R., Nesdill D., Park C., Reecy J., Reiser L., Sanderson L.A., Sen T.Z., Staton M., Subramaniam S., Tello-Ruiz M.K., Unda V., Unni D., Wang L., Ware D., Wegrzyn J., Williams J., Woodhouse M., *et al.*, 2018. AgBioData consortium recommendations for sustainable genomics and genetics databases for agriculture. Database (Oxford), 2018.

iKC, 2013. The i5K Initiative: advancing arthropod genomics for knowledge, human health, agriculture, and the environment. J Hered 104: 595-600.

McCarthy F.M., Wang N., Magee G.B., Nanduri B., Lawrence M.L., Camon E.B., Barrell D.G., Hill D.P., Dolan M.E., Williams W.P., Luthe D.S., Bridges S.M., Burgess S.C., 2006. AgBase: a functional genomics resource for agriculture. BMC Genomics 7: 229.

Mullet J.E., Klein R.R., Klein P.E., 2002. Sorghum bicolor - An important species for comparative grass genomics and a source of beneficial genes for agriculture. Curr Opin Plant Biol 5: 118-121.

Paul I., Bhadoria P.S., Mitra A., 2016. Plant Volatile Genomics: Recent Developments and Putative Applications in Agriculture. Recent Pat Biotechnol 10: 4-11.

Rothschild M.F., Plastow G.S., 2008. Impact of genomics on animal agriculture and opportunities for animal health. Trends Biotechnol 26: 21-25.

Stokes M.E., McCourt, P., 2014. Towards personalized agriculture: what chemical genomics can bring to plant biotechnology. Front Plant Sci 5: 344.

Studholme D.J., Harris B., Le Cocq K., Winsbury R., Perera V., Ryder L., Ward J.L., Beale M.H., Thornton C.R., Grant M., 2013. Investigating the beneficial traits of Trichoderma hamatum GD12 for sustainable agriculture-insights from genomics. Front Plant Sci 4: 258.

Varshney R.K., Thudi M., Pandey M.K., Tardieu F., Ojiewo C., Vadez V., Whitbread A.M., Siddique K.H.M., Nguyen H.T., Carberry P.S., Bergvinson D., 2018. Accelerating genetic gains in legumes for the development of prosperous smallholder agriculture: integrating genomics, phenotyping, systems modelling and agronomy. J Exp Bot 69: 3293-3312.

Wang W., Cao X.H., Miclaus M., Xu J., Xiong W., 2017. The Promise of Agriculture Genomics. Int J Genomics 2017: 9743-749.

□□□

Chapter **12**

Untold Complex Constrains in the Startups for Bio-fuel Promotion – A Live Experience

K. Narayana Gowda

Former Project Coordinator, RBRC Project and Former VC, UAS, Bengaluru-560065
E-mail: knarayanagowda@yahoo.co.in

Rural Bio-Resource Complex Project (RBRC) funded by Department of Biotechnology, Government of India was implemented by UAS-B during April 2005 to March 2010 involving 8340 families spread over in 75 villages of Tubagere hobli in Doddaballapur taluk, Bengaluru rural district. The main objective was to bring in overall improvement in the standard of living of all the families which includes both landed and landless people more specifically, farmers, women, youth, besides local artisans, farm labourers and even physically challenged people with a concern to bring in inclusive growth and sustainable development.

Initially, 24 interventions were deliberately chosen keeping the natural resources, family requirement, market demand, knowledge and skills besides local interests and culture to meet all sections of the society. Value addition and processing was given top most priority with the intention of enhancing income and create more employment opportunities. It was to bring in increased productivity, higher employment opportunities, augmenting significant income generation leading to very satisfying living of all families 1.

Among the 24 interventions identified by the project team, Bio-fuel promotion was one of the interventions taken up under the project. Comprehensive empowerment activities were organised including campaigns to promote large scale bio-fuel plantation. The objective was to bring non-arable and poorly used land both, in private and the public holdings under bio-fuel plantation to improve stability of income, environment and employment generation.

Initially, in the first year, Bio-fuel seedlings available with the University and State Department of Forestry were purchased and made available to the potential farmers at free of cost during the appropriate season.

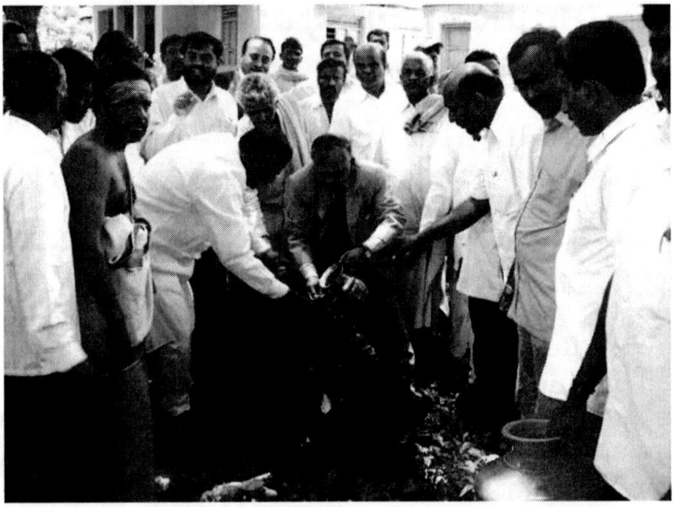

Inauguration of Biofuel Planting Campaign by Dr.P.G.Chengappa, the then Vice-Chancellor, UAS, Bangalore at Melekote Village of Doddaballapur Taluk

Innovative Mechanism Evolved for Raising Bio-fuel Seedlings

The cost of each seedling including transport and handling used to be around Rs. 10. Thereafter, with the debate and discussions, it was decided to train the youth and entrust responsibility of raising bio-fuel seedlings the under-employed youth during the beginning of second year. Altogether from five panchayats, 25 interested youth were trained for two days at the GKVK campus of the University by two Professors of Forestry Department, who were actively involved in the project. Since the technology was simple and the youth were formed into small groups of 3 to 4 from each panchayat and provided guidance at the village level, to raise timely and quality seedlings.

An innovative mechanism was evolved to ensure the availability of seedlings at the easy reach as well as timely to every farmer. The cost of raising the seedling used to be Rs.3 per seedling and Rs. 8 per seedling was paid to the youth, which saved Rs. 2 per seedling to the project. Later these youths were smart enough to go for raising high value seedlings/grafts, such as Mango and Jack fruit. The survey conducted after two years of plantation (2008-09) revealed that the percentage of survival of seedlings was 94%. The major reason being, seedlings were planted at the appropriate time at the leisure time of farmer and when there was enough moisture in the soil. Large area under Bio-fuel plantation was taken up even by school children, who evinced lot of interest to take up plantation under the guidance of teachers.

Preparation of Grafts of Bio-fuel Species by Youths in the Project Area

Involvement of School Children, Teachers and Farmers in Bio-fuel Species Plantation Programme

Initially, project team heavily relied on timely market intelligence information for the farmers to avail better price. But it was realised at the end of first year of the project implementation that market intelligence information was mainly useful to farmers producing high-value crops that too in large quantities. Since more than 90% of farmers were small and marginal, majority of them engaged in low value crops producing smaller salable quantities, the major part of their farm produce sold to middleman at a very low price, market intelligence information strategy as conceived in the beginning of the project, was not an appropriate solution to all farm produces and farmers.

Data analysis conducted revealed that the farmers share was less than 30 per cent, out of consumer share across many farm produces and in many cases, it was less than 20 per cent. During the second year, a series of discussions started on marketing of all farm produces. Marketing of Bio-fuel seeds was found to be in the hands of middlemen for decades. Pongamia seeds were paid Rs.5 per kg and Neem Rs.8 per kg during 2006. Therefore, the farmers were keen on getting a respectable price for the bio-fuel seeds.

Genesis of Organised Group

A series of meetings, discussions, debates and experience sharing activities were started both, at the University level and project area to find sustainable measures for marketing. After a series of meetings and consultations with marketing experts, it was felt appropriate to organise the farmers into meaningful groups so that they can stand together to address marketing. A variety of options were discussed weighing the pros and cons of organised group because by then, there was no standardised procedure for procurement and marketing.

Specific to Bio-fuels marketing, separate meetings were held and thereafter, it was agreed to start Rural Biofuel Growers Association (RBGA) in the project area to address procurement, processing and marketing. The Research Fellows working in the project area were advised to enrol the Members for the Rural Bio-fuel Growers Association by collecting a membership fee of Rs.100 per family. After a minimum of 60 members' enrolment, office bearers were formed and it was decided to register the Association. Neither the office bearers of the association nor the project team was knowledgeable about the registration procedure. After a series of consultations with all the concerned authorities (even the concerned officials were not familiar since there was no such Bio-fuel Association in their records), the Association was registered with the Registrar of Firms and Societies during 2008. Left to the office bearers, it would have taken a couple of months for registration while the project team completed the task with in a fortnight.

Thereafter, a series of meetings were organised involving all the stakeholders on the site for establishing a processing plant, preparing project proposal on the budget required, type of machineries, accessories, manpower, building, procurement, payment, etc. There was no readymade information available for addressing the aforesaid issues anywhere in the country except some small scale oil extraction plants established by private entrepreneurs. The project report was prepared after six rounds of meetings and discussions involving stakeholders under the leadership of the project team. Establishing a processing plant under organised group is more challenging and time consuming, when compared to the individual entrepreneur.

Location of the Processing Plant

Representatives from each panchayat were demanding that the processing plant to be located in their panchayats jurisdiction and nobody was prepared to compromise. It was resolved after two rounds of meetings keeping requirements of certain minimum infrastructure facilities, such as road, access for large number of visitors, proximity to the village, etc., for the ideal location of the plant.

Acquiring Land for Establishing Processing Plant

The concerned panchayat agreed to hand over 20 gunthas (around 20,000 sft) of land for the purpose after two special meetings held for taking decision. Accordingly, a letter was addressed by the panchayat Secretary to the Tehsildar to hand over the said land to the Association for the purpose. But the Tehsildar blindly refused stating that there is no government order to hand over land for such purpose. Even the appeal made to Assistant Commissioner of the jurisdictional area did not solve the problem. The project team was unhappy about the attitude of the Revenue Department officials. Thereafter, the team met the Deputy Commissioner of the district and appraised the importance of the project. Deputy Commissioner was convinced of the facts and ordered for the sanction of land for this purpose. Left to the local people and office bearers, the issue might have taken a couple of months but the project staff got it within a fortnight without any aditional expenditure.

Budget

The budget required for building, machineries, initial expenditure, etc., was got estimated with a team of experienced personnel and it was Rs.11 lakhs.

Interestingly, the membership amount realised was only Rs.6000 collected from 60 members in a span of six months of intensive efforts. The response to the membership drive was very poor and all the Research Fellows were unhappy about this task, at one stage they were expressing that they are prepared to pay from their pocket than pursuing the farmers to become members. The fact is that the farmers are unwilling to become members for a low tech enterprise compared to Hi-tech enterprise, such as Grape growers, Pomegranate growers where in the volunteer to become members.

The question before the project team was how to mobilise huge amount of Rs.11 lakhs for the purpose. Series of meetings and discussions were held on mobilising finance for the purpose. The project team agreed to provide Rs.2 lakhs on the condition that the office bearers and local institutions would contribute matching amount. Most of the Directors wanted to resign and go away from the association, when this proposal was putforth but because of the good wisdom prevailed among few local leaders; they agreed to contribute the matching amount. Thereafter, each Director paid Rs.10,000 and all the twelve of them paid Rs 1,20,000. Forty nine Milk Producers Co-operative Societies (MPCS) Secretaries were convinced to pay Rs. 2000 after several rounds of meetings realising an amount of Rs. 98,000. The Department of Agriculture was approached with the proposal, which realised a sum of Rs. 50,000. Despite all these efforts, there was a shortfall of nearly Rs.6 lakhs to up the project proposal.

Later, a decision was taken to approach the Corporation Bank for obtaining bank loan for the purpose. The local Corporation Bank Manager was contacted to provide the loan. The Manager plainly refused stating that there is no such provision in the banking rules to provide loan for Associations despite the fact that earlier the loan was provided to other enterprises proposed by the Project Team and familiar about the project activities. Further, observation of the Manager was that office bearers quite often changes and who would take the responsibility of repayment.

Thereafter, the project team pursued the issue with the Regional Office of the Corporation Bank. Assistant General Manager of the Bank (happens to be the classmate of the Project Co-ordinator) told that we go with the Branch Manager, although some efforts were made to convince the Branch Manager which was futile. Later, efforts were made to convince the Chairman of the Corporation Bank, who agreed to speak to the Manager to consider this as a special case. The Manager unwillingly agreed to provide the loan stipulating many conditions. Although Project Team was able to fulfill majority of the conditions put forth by the Manager, the real problem started with giving security for the loan availed. Since it is an association, Manager was asking who should give the security. Although the office bearers agreed to give security, it was turned down since office bearership is a tenural responsibility and accountability cannot be fixed on them. The Bank Manager was asking for government employee to provide security and insisted that the loan will be sanctioned, if the condition was fulfilled.

This condition was neither comfortable to the project staff nor the office bearers. There was a week long discussion on how to solve this complex problem and majority want to drop the proposal for establishing the processing plant. The reason being the office bearers were not government employees while all the project staff was temporary except the Project Co-ordinator and Associate Project Co-ordinator. As a project leader, it was unpalatable and embarrassing issue to ask the Associate Project Co-ordinator to provide security for the purpose. At the end, Project Co-ordinator volunteered to offer security for this purpose. Thus, the funds required for the project was solved with great deal of efforts. Even the Bank Manager was astonished to know the commitment and concern of the Project Co-ordinator. This was the most unexpected complex untold situation the project staff faced in the implementation of the project, although there were many such untold challenges but they were relatively of lesser magnitude.

Procurement of Machinery

Initially, it was planned to procure machinery from either Japan or China in view of high-quality machine being seen in one of the International Bio-fuel Conference by the project team in Spain. But the foreign policy involved in procurement was not only cumbersome but also time-consuming. Therefore, it was proposed to purchase from

Coimbatore Company after a thorough discussion with engineers specialised in the area keeping in view quality and cost factors. Three of the office bearers were deputed to Coimbatore to train on the operation and maintenance of machinery.

Procurement Mechanism of Bio-fuel Seeds

Major cost involved in the sale of farm produces is transport and handling. This is more, so, when suitable produce will be a smaller quantity and to be taken to far-off places for marketing. Bio-fuel produce is one such produce. Therefore, series of debates and discussions were held involving concerned stakeholders. Project team suggested the noble idea of procurement through MPCSs. Incidentally, milk societies work two hours in the morning and another two hours in the evening. Prime hours between 10 AM to 4 PM, the manpower, machineries and building were underused was proposed for procurement and payment. The purpose was well-received by all the farmers as well as the office bearers which were operationalized after three rounds of meetings with Secretaries of MPCSs and their higher-ups. It was agreed to pay 5 per cent of transactions to the MPCS, where in 3 per cent to the Secretary and 2 per cent to the MPCS.

Co-operation, collaboration and co-ordination are being discussed heavily at the top level but fail most of the times at the grassroot level. It is possible to ensure good collaboration and co-ordination, if suitable empowerment, recognition and encouragement are provided to personnel working at different levels in the system.

Although the system was used for bio-fuels, later on it was used by the project staff for jackfruit and other minor produces procurement including the distribution of critical inputs. This was yet another innovative mechanism, the project evolved. Karnataka state has more than 14,500 MPCSs, while the country has 1,64,000 MPCSs connecting most of the villages across the country. The system with some more refinement will bring in reducing overhead costs by distributing inputs as well as procurement of outputs improving substantial profit margin to farmers and wide range of other benefits.

The construction of buildings, procurement and installation was done during 2008 as per the plan. The functioning of machinery was a difficult proposition since the electricity available in the area was insufficient to run the machine installed. The Taluk Electricity Board officials were unwilling to provide transformer to the plant site. The project staff contacted State Electricity Authorities and convinced the importance of Bio-fuels in generating renewable energy. State authorities positively responded to the request and sanctioned additional transformer, which enabled efficient functioning of the machinery installed, yet another great support by the project staff left to the office bearers, it would have been herculean task.

Three young unemployed youth were appointed and trained by machinery suppliers in the presence of office bearers on operation and maintenance. Ultimately, the plant was commissioned during January 2009, which took more than two years with 42 rounds of meetings, visits and training programs. The rate to be fixed for the bio-fuel seeds was also discussed involving all the stakeholders including economists on cost-benefit analysis. Finally, it was decided to pay Rs. 15 per kg for Pongamia and Neem at Rs. 20 per kg. The sale of Pongamia oil at Rs. 50, while neem oil at Rs. 60, where cake at Rs. 50 and Rs. 20, respectively. The rates were fixed in such a way to take care of bank EMI payment, wages, expenses of office maintenance and miscellaneous expenditures besides saving for expansion and maintenance.

The supply of bio-fuels seeds was done by not only bio-fuel farmers but also by SHGs and school going children which created indirect employment to many, particularly during off season. Bio-fuels will come for harvest during summer season and bio-fuels are generally grown in rainfed areas. Although the anticipated yield was 120 tonnes but interestingly, it was going at more than 180 tonnes during 2009. The reasons observed for increased realisation of yield were: profitable price, procurement at the village level through MPCS, transparency in weighment and timely payment which was going unharvested.

In summary, left to the people the establishment of RBGA was only a dream come true. Similarly, without the Bio-fuel growers' effective and active participation, the establishment of RBGA would not have been possible and continuation of RBGA activities during the post-project period would have been difficult. The establishment of KVK during 2008 by ICAR under UAS (B) in the project area was a great advantage in providing follow up guidance and additional support. Demand for cake so much that the Secretary of RBGA used to mention it as ATM that many hi-tech farmers wait in queue for three months by paying advance. The demand for Neem oil was extremely good but the sale of Pongamia oil was a difficult task. During 2010, the establishment of Bio-fuel Esterification Unit with the financial support from Karnataka Bio-fuel Development Board, which was facilitated by KVK to process Pongamia oil into bio-diesel came handy and there was a huge demand for bio-diesel oil. The bio-diesel oil was used for University heavy vehicles at 20 per cent mix besides KSRTC. Thereafter, the demand for both the products started increasing. Many farmers from outside the project area started taking advantage of the system evolved. But the repayment of loan availed was only rupees two lakhs and balance of rupees four lakhs was still due during 2012. Then, the Vice-Chancellor arranged to pay the balance of rupees four lakhs with RKVY funds.

Currently, the RBGA is functioning extremely well on its own with limited dependency from KVK and other institutions. The procurement of Pongamia seeds @ Rs. 32 per kg and Rs. 40 per kg for Neem seeds is being done. The model was the centre of attraction for academicians, planners, administrators, extension workers, NGOs, students and farmers from the state, country and even abroad. Then DG, ICAR, Dr S.Ayyappan, Governor of Karnataka, Dr Hansraj Bhardwaj, former Prime Minister of India, Shri H.D. Deve Gowda and other important dignitaries visited the centre and interacted with office bearers and bio-fuel growers. Everyone expressed that this is the best model that can be replicated across the country and even in contemporary countries.

Visit of Sri H.D.Deve Gowda, Former Prime Minister & MP to Rural Bio-fuel Processing Unit at Hadonahally, Doddaballapur Taluk

Inauguration of Rural Bio-fuel Processing Unit at Hadonahally Doddaballapur Taluk by the then Governor of Karnataka Dr. Hansaraj Bhardwaj

Visit of Dr.S.Ayyappan, Secretary, DARE & DG, ICAR to Rural Bio-fuel Processing Unit at Hadonahally, Doddaballapur Taluk

India has great advantage in bringing sizeable bio-fuel area under non-arable land, which is left barren and underutilised. This was multi-farious advantages in improving overall economy of the country, particularly improving and stabilising the income of farmers in low rainfall region and minimising the import of fossil fuel oil, provides employment to large number of unemployed and under-employed youth above all improving soil health and empowerment.

Conclusion and Recommendations

There is absolute need for harnessing Bio-fuel resources by availing the sustainable and an end to end model evolved by UAS, Bengaluru. The responsibility of planning and implementation may be entrusted to the Department of Agriculture in close co-ordination with respective KVKs and NGOs in the area. The Centre and State governments to earmark enough separate budget for the purpose. The procedures should be pro-farmer and enough leverages to be provided, particularly single window system to speed up the implementation. The security for availing loan from Banks for such purposes should be liberal with higher margin of grants. The esterification unit may be established linking every 5 to 10 processing units, which will become economically viable and sustainable.

References

Narayana Gowda, K., 2011, *DBT-Rural Bio-Resource Complex Project Report* (2005-06 to 2009-10). University of Agricultural Sciences, Bengaluru, P. 112.

Narayana Gowda, K., 2012, *Bio-fuel Growers Market Network – A Success Story*. Asia-Pacific Association of Agricultural Research Institutions, FAO Regional Office for Asia and Pacific, Bangkok, Thailand, 30.

□□□

Chapter 13

Opportunities and Challenges for Agri-entrepreneurships in Promotion of Biological Control for Crop and Soil Health - A Case Study with Entomopathogenic Nematodes

M. Nagesh[1], Letha Devi G[2] and Chandish R Ballal[1]

[1]ICAR–NBAIR and [2]ICAR - NIANP, Bengaluru, E-mail:nagesh.m@icar.gov.in

Smart agriculture broadly envisages worldwide linking of information flow, create temporal and spatial opportunities in agriculture and providedecision systems for the benefit of stakeholders, especially, the farming community. It is essentially pivoted on the ideology of demand-supply, innovation-agribusiness, and an agriculture-a profitable proposition to all stake holders. Smart agriculture essentially requires innovations, technologies, investment, skill development, agripreneurs, strategies and stakeholders. Agri-startups have a great potential to link the lab technologies to end-users and agripreneurs to investment, farm produce to market chain, production to demand-supply, investment to employment generation and profitability to farmers.

Among several prospective propositions available for agri-entrepreneurship as start-ups, establishment of input production and supply chain in farm sector are gaining prominence. In this presentation the authors attempted to discuss the role of the ICAR-National Bureau of Agricultural Insect Resources (erstwhile Project Directorate of Biological Control), Bengaluru, in developing several agripreneurships in the field of biological control of crop pests and diseases, through IP-driven technologies and innovations.

Management of inputs for environmentally-safe plant and soil health protection through biological means (biological control) is a growing component of Smart Agriculture and organic farming. ICAR-National Bureau of Agricultural Insect Resources (erstwhile Project Directorate of Biological Control), Bengaluru has been striving hard to catalogue the biodiversity of beneficial flora and fuana related to agri-pests, diseases and weeds; evolve their mass production and formulation technologies; validate their field efficacy against several pests, andevolve technologies/innovations as per the compliances of Central Insecticides Board and Registration Committee (CIB & RC), National Biodiversity Authority (NBA) and other international regulations.Several innovations werepatented and licensed to agripreneurs promoting effective mentoring and handholding of agri-startups in the field of input supply for biological control of crop pests (Table 1).

Table 1: Technologies/products developed by ICAR-NBAIR for biological control of crop pests and available for agri-entreprenuership

S. No.	Title
1.	Multiple insecticide tolerant strain of egg parasitoid, *Trichogrammachilonis*
2.	High temperature tolerant strain of egg parasitoid *Trichogrammachilonis*
3.	Pesticide tolerant strain of aphid lion, *Chrysoperlazastrowisillemi*, an important predator of sucking pests
4.	Novel insecticidal WP formulations of *Heterorhabditisindica* for the biological control of white grubs & other soil insect pests
5.	Novel wettable powder formulation of *Pochoniachlamydosporia* as bionematicide against plant parasitic nematodes
6.	Liquid formulation of *Bacillus thuringiensis*
7.	Powder based formulation of *Pseudomonas fluroscens*, a DAPG producing abiotic stress tolerant isolate for rainfed and stressed agricultural soil
8.	Closed system for mass production of predatory mites
9.	A dispenser for the monitoring of eucalyptus gall wasp
10.	Bio formulation of salinity tolerant *Trichodermaharzianum* with biocontrol potential
11.	Bioformulation of carbendazim tolerant *Trichodermaharzianum* with biocontrol potential
12.	Powder based formulation of *Bacillus megaterium* as growth promoter & management of bacterial wilt disease
13.	A plant volatile based attractant for enhanced attraction of fruit fly
14.	A simple technique of rearing brinjal shoot and fruit borer, *Leucinodesorbonalis*

15. Protocol for designing lure for impregnating parapheromone 4[4-acetoxy] phenyl-butanone to attract male flies of *Bactrocera* spp attacking cucurbit crops for mass trapping and monitoring its population thereof
16. Control release dispensers for semiochemicals
17. Mass production of *Trichogrammachilonis & T. embryophagum* using Eri silkworm eggs
18. A Herbal based Repellant for Termites on woody trees
19. A. Herbal swabber for the management of white stem borer *Xylotrechusquadripes* in Coffee (organic and non-pesticidal).
20. B. Booster for boosting plant health in coffee (not for certified organic coffee)
21. Adsorption and delivery of molecules using Nanoporous materials
22. Dorsa-Delta, an efficient trap for mango fruit fly
23. A Bisexual attractant for *Bactroceradorsalis* in delta trap
24. A Technique for rearing of Housefly parasitoid *Spalangia*
25. A Technique for rearing of Housefly parasitoid *Nasoniavitripennis* (Pteromalidae)
26. Waste to wealth: Technology on Black Soldier Fly mediated bioconversion of farm and kitchen wastes
27. Insect repellent formulation and methods thereof
28. Novel Device for field release of parasitoids

In this backdrop, we wish to share our experiences in the form of a case study on the prospects of agri-entrepreneurship in biological control, the constraints and some suggested leads for way forward.The case study refers to the technologies relating to soil-dwelling beneficial organisms, i.e., entomopathogenic nematodes (EPN) for biological pest management byICAR-NBAIRand developthem in to agri-business proposition, including IP protection.

Globally, 4 decades agoEPN were well-recognized as effective and commercially viable biocontrol agents for pest management suiting to agribusiness leading to availability of EPN. During late 80s, M/S Ecomax, Hyderabadmade the first bold commercial venture to import and supply EPN in aqueous formulations, which could not sustain in the market and was completely recalled from the business.Significantly, the factors contributing to the failure were, the M/S Ecomax imported without production systems established in India;limited awareness and demand among farming communities and researchers about the product (EPN);ecological suitability of exotic strains of EPN;limited shelf-life and unsuitable formulation; lack of validation in field against pests, etc. Subsequently during the next 3 decades, research, awareness and innovations rapidly increased world-wide and pan India. At present EPN are available as commercial products across the board in Americas, Europe, Australia, China, Japan etc.

During last 2 decades, ICAR-NBAIR successfully addressed the challenge of evolving EPN technologies and transforming them in to commercially viable propositions. The in-house support from NBAIR and ICAR, and grants from DBT and DST, thankfully acknowledged, which helped us to systematically catalogue native diversity of EPN, screen their bioefficacy against soil-borne crop pests, evolve POPs, and patent technologies for their scale-up production, formulation and field use. Innovations in scale-up *in vivo* production and formulation with 12 months shelf-life, were filed for patent in 2010 and were granted IP rights in 2018. During the last 8 years (2010-18), we licensed the know-how to 28 commercial entrepreneurs and technically mentored them to achieve an annual production of 1000 metric tonnes (Fig. 1). An area of about 30,000 ha is currently under EPN (Fig. 2) in different crops of economic significance (Fig. 3), in different states (Fig. 4), mitigating use of soil insecticides and contamination to soil and water bodies.

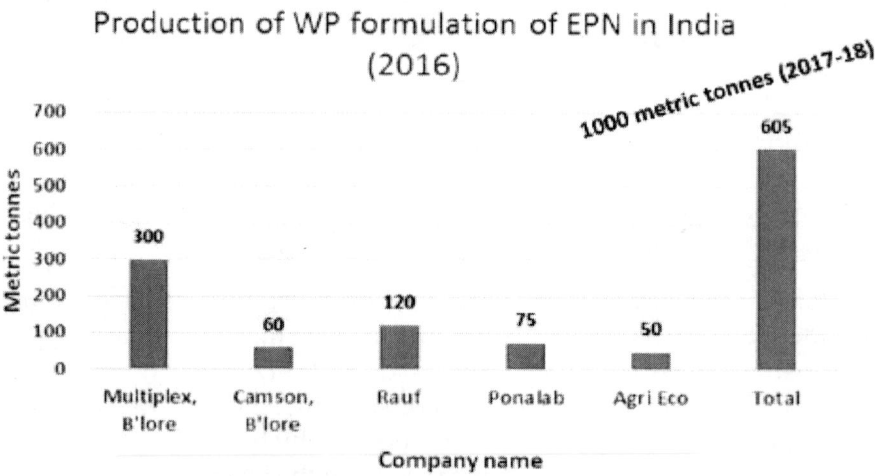

Fig. 1: Indigenous production of EPN by Indian entrepreneurs promoted by ICAR-NBAIR.

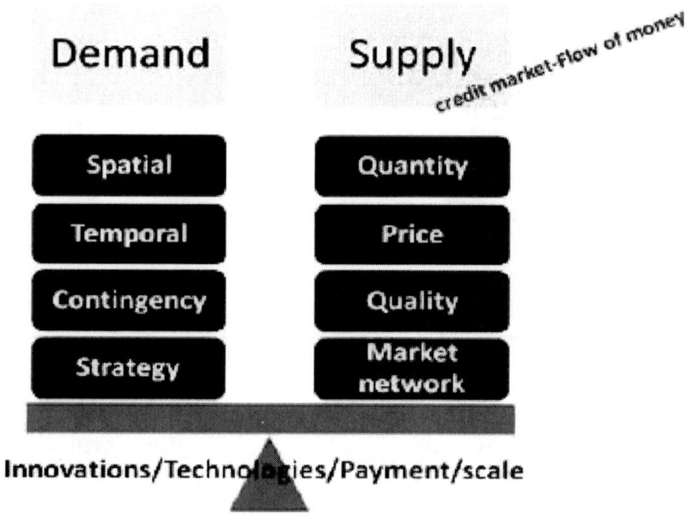

Fig. 5: Factors balancing demand-supply situation for agri-inputs including biocontrol agents.

Estimated annual demand for EPN for the management of soil-borne insect pests including whitegrubs, termites, root grubs, cut worms, army worms etc., primarily in crops including cardamom, banana, groundnut, sugarcane, arecanut and vegetables amount to 24,000 metric tonnes (Nagesh*et al.*, 2017). ICAR-NBAIR in its efforts to address the availability of the EPN product to the farmers, licensed the patented technology to as many entrepreneurs as possible on non-exclusive basis on PPP mode (Table 2). The current production levels (based on estimations and inputs available) are 1,000 metric tonnes during 2017-18 (Fig. 1), with an adoption area of 30,000ha (2018) (Fig. 2). Among the crops in which EPN technology is adopted, sugarcane, groundnut, arecanut, cardamom, banana and brinjal are prominent (Fig. 3), while state-wise adoption of the technology indicates that Maharashtra followed by Karnataka, Kerala, Tamil Nadu, Gujarat, Rajasthan, Madhya Pradesh and NEH regions are primary (Fig. 4).

ii) Demand and agri-business proposition on EPN

Estimated demand across the country is 24,000 metric tonnes of WP formulations of EPN to cover an estimated area of 5,00,000 ha against 18-20 major pests in 14 crops (Table 3). Our efforts on EPN technology encouraged and promoted the ideologies of Make-in-India startups in agribusiness. However, the production-supply chain for

EPN is in infancy with current estimated production 0f 1,000 tonnes. These estimated figures clearly indicate that there is a vast scope for agri-business to be established and for stakeholders. Current production systems are based on *in vivo* technologies with an inherent limitation of scalability. Scalability can be achieved either by bringing-inmore number of agri-preneurs, switching over to high-investment*in vitro* production systems or a combination of both.

Table 3: Conservative estimate of factors that contribute to demand and agri-business proposition on EPN (2017-18)

Factors	Estimate	Consideration
Area to be covered	5,00,000 ha	2 seasons
Number of pests to cover	18-20	Root grubs, cutworms, borers, termites, mole crickets
Number of crops	>14	Sugarcane, groundnut, arecanut, banana, ginger, turmeric, fruit, cardamom, oilseeds
Requirement	24000 tons WP	
Production cap. (*In vivo*) by 8 active companies	1000 tons p.a.	*In vitro*
Number of companies	50 or more	Scope for market

Other important considerations for agri-business start-ups is initial investment/capital costs, gestation period and Returns-on-Investment (ROI). An initial investment for a star-up in EPN technology is at Rs. 15 lakhs per unit, with agestation period of 3-4 years for commercially high ROI.

Furthermore, the health of an agri-business is dependent not only on temporal and spatial supply but also on the commitment to quality assurance, prices, flow of sale proceedings, biosafety standards etc., after sale customer satisfaction.

Experience gained and analysis of behavior of agri-entrepreneurship related to EPN

During the last 10 years of licensing of EPN technologies and promotion of entrepreneurship among about 28 companies across the country, we addressed challenging queries from the industry w.r.t. scale-up production chains, investments, agri-supply chain, bulk and seasonal market, transport, storage, quality standards, flow of sale proceedings, registration and biosafety standards.Noteworthy, EPN technology is ecologically safe and not notifiedunder registration requirements of CIB and RC. Like any natural predator and parasitoid, EPN are natural parasites of insect pest species and do not feed on any beneficial soil fauna, fish, birds, reptiles, mammals or human being. Despite the efforts on innovations, handholding, enthusiasm from the entreprenuers etc., there has been a stagnation in growth of product availability, quality and decline in active producers. The major fallout observed is, a monopolistic trend with high pricing and low compliance of quality standards, which is most likely lead to a positive spin off of more number of entrepreneurs or a negative trend of low field performance of the product in the field and gradual rejection in the market.

We analyzed our experiences in EPN agri-entrepreneurship of last 10 years and presented the summary in Table 4. We examined a total 24 licensees of our EPN technologies till 2017 for various factors as listed in Table 4, and the significant observations are as follows:

- The licensees can be categorized in to 4 - No starters; Mid-show quitters; Prospective players and Main players.
- No starters are licensees who have taken the technology but have not really ventured in to the production mainly due to intricacies involved in production technologies, lack of awareness of various factors involved in production to end-user through market network, lack of production and market infrastructure, etc. Some of the non-starters have invested on EPN technology because there was seed money/external funding available for this purpose, through BIRAC, DBT, DST etc. Some of the non-starter licensees wanted a very short gestation of 9-10 months for scale-up. Their percentage was about 25%.
- Mid-show quitters among the licensees were 21%. They are enthusiasticstarters, with high expectations on ROI in short gestation period, with or without institutional financial support, reasonably good awareness of the technology and market, low inhouse and market network infrastructure and with intention to diversify than specialize. Though the motivation was due to the demand for EPN in the market, the quality and gestation requirements are inhibitory.
- The 3rd category, Prospective players are about 33% of the licensees, who are characteristically the investors motivated by the demand for EPN, awareness and faith in the technology,

and acceptance to the quality and gestation requirements. They are either group investors or players who are in to other products earlier, with medium risk bearing ability, in-house infrastructure and market network.

- The 4th category, the Main players, are to the tune of 21%, who have good knowledge of technology, vivid motivation to meet the demand of farmers, inhouse and market network capabilities to meet the spatial and temporal demand. These are licensees with group, corporate or company funding with high risk bearing ability.
- Scale-up infrastructure, gestation, fund source, market network and motivation to meet the demand were important pointers for the agri-business on EPN.
- Due to imbalance in supply-demand-producers there has been a negative effect of monopoly on quality and pricing of the product. Companies with better market network capitalized the demand and brand name affecting the price and quality.
- More importantly, some of the licensees expressed that the input supply chain has payment-credit situation, and recovery of the payments due from the endusers is a longstanding limitation.

The way forward

The trends in general agribusiness and agri-entrepreneurship are very encouraging, ascending and sustaining. The demand and supply of biocontrol agents have undoubtedly opened new avenues for agribusiness in a big way. The main driving factors include growing awareness among stakeholders (farmers, consumers, policy makers, environmentalists, health care systems, international regulations) about the necessity to comply with ecologically safe, residue-free, non-contaminating and biologically safe technologies/innovations for crop pest management, and gradual minimal or judicious use of synthetics. Although, there are inherent constraints attached to these agri-inputs (biocontrol agents) in terms of niche market, seasonality, bulky volumes, short shelflife, fluctuating prices etc., these factors are also opportunities for technological innovations, market intelligence, strong supply chains, bulk storing and supply networks, etc. Due to demand-supply As observed, interestingly there has been a gradual change in outlook and complexion of the investors, i.e., collective funding, corporate and venture capitalists are investing in production and supply of bioagents for agricultural use. Till recently, it was either a kind of cottage industry or a small inhouse business proposition, which needs a paradigm transformation in investment, institutional support and net working to meet the farm needs.

The authors arrived at the following points to strengthen and promote agri-preneurship:

- Transformation in a *lifestyle business-startup* and *small business-startup to scalable startup* strongly supported by technological innovations in promotion of biological control as agribusiness. Agri-innovation centres of national institutes, SAUs, NARES etc., should have business plans developed, for the technologies that were generated by them. Handholding and mentoring by the innovators for the licensed technologies would immensely help in providing the much needed faith and confidence in agri-business.
- Establish a tangible input (biocontrol)-supply chain mechanism which will have an accountability on quality, price-line, timely supply, payment flow and recovery of payments to the producers/agripreneurs. The Farmer Produce Organizations (FPOs), cooperative societies, etc., are now emerging as very effective self-help institutions. There is an assured and direct benefit to the agri-startups to get networked with these FPOs, besides directly benefitting the farmers and the existence of FPOs.
- Agri-startups have a wide scope in quality certification and regulation,mobile apps linking needy farmers, FPOs, financial institutions, technical staff, input suppliers, bulk storage facilities, market etc., facilitating flow of spatial and temporal information.
- Institutional support for finance and refinance for agri-startups through PPP, BIRAC, NABARD, FPOs etc., would help enterprising qualified youth to energize the production-supply chain networking.

It is summarized that the new policy initiatives supported aptly by NABARD, NHB, NAFED, FPOs, investors, are now focusing onstreamlining the innovations-production-storage-supply-marketing-investment for promotion of organized agri-business, income and employment generation, better livelihood, resource use and trusteeship of natural resources.

Table 4: Analysis of status of technology transfer/license to entrepreneurs and the inherent problems

Pointers	No Starters	Mid Show Quits	Prospective Players	Main Players
No. of licensees (24)	6 (25%)	5 (21%)	8 (33%)	5 (21%)
Prior knowledge/skill	Nil	Nil or little	Yes	Yes
Faith in technology & gestation period	Start only	Flair for short gestation, scaleup but not start from low scale.	Yes	Yes
Investment/seed funding	Limited/sourced	Medium/sourced	Medium/self/group	High/company
Risk taking ability	Limited	Medium	Medium	High
Market network	Limited (few locations)	Medium (few districts in local state, personalized)	Low to medium (distributors)	High (country-wide)
Diversification	6-8 other products	8-10 other products	< 6 other products	> 10 other products
Institutional support	BIRAC, State Science Council, DST, DBT, SAU	BIRAC, State Science Council, DST, DBT	NIL	Nil
Quality status	-	Medium	Very good	Poor-medium; Demand-supply-affluence oriented
Price status	-	Medium to high	Medium	3-4 times higher, Monopoly
Product supply status	-	Low	Low	Monopoly, quality compliance
Motivation for startup	Funding by govt, demand for EPN	Demand for EPN	Demand for EPN	Profitability, brand name
Suggested action	Auditing of Public funding agencies	Motivation, hand holding, Innovations, market support	Institutional fund support, Handholding, Innovations, market support	Regulatory measures